T0201284

Developments in Strategic Materials and Computational Design II

Developments in Strategic Materials and Computational Design II

A Collection of Papers Presented at the 35th International Conference on Advanced Ceramics and Composites January 23–28, 2011 Daytona Beach, Florida

Edited by
Waltraud M. Kriven
Andrew L. Gyekenyesi
Jingyang Wang

Volume Editors
Sujanto Widjaja
Dileep Singh

A John Wiley & Sons, Inc., Publication

Published by John Wiley & Sons, Inc., Hoboken, New Jersey.
Published simultaneously in Canada.

For general information on our other products and services or for technical support, please contact our
Customer Care Department within the United States at (800) 762-2974, outside the United States at
(317) 572-3993 or fax (317) 572-4002.

Wiley also publishes its books in a variety of electronic formats. Some content that appears in print may
not be available in electronic formats. For more information about Wiley products, visit our web site at
www.wiley.com.

Library of Congress Cataloging-in-Publication Data is available.

ISBN 978-1-118-05995-1

oBook ISBN: 978-1-118-09539-3
ePDF ISBN: 978-1-118-17237-7

ISSN: 0196-6219

Printed in the United States of America.

10 9 8 7 6 5 4 3 2 1

Contents

THERMAL MANAGEMENT MATERIALS AND TECHNOLOGIES

COMPUTATIONAL DESIGN

ADVANCED SENSOR TECHNOLOGY

Preface

Contributions from two Symposia and two Focused Sessions that were part of the 35th International Conference on Advanced Ceramics and Composites (ICACC), in Daytona Beach, FL, January 23–28, 2011 are presented in this volume. The broad range of topics is captured by the Symposia and Focused Session titles, which are listed as follows: Focused Session 1 - Geopolymers and other Inorganic Polymers; Focused Session 2 - Computational Design, Modeling Simulation and Characterization of Ceramics and Composites; Symposium 10—Thermal Management Materials and Technologies; and Symposium 11—Advanced Sensor Technology.

Focused Session 1 on Geopolymers and other Inorganic Polymers was the 9th continuous year that it was held. It continues to attract growing attention from international researchers (USA, Belgium, Jordan, Germany, France, Czech Republic, and Greece) and it is encouraging to see the variety of established and new applications being found for these novel and potentially useful materials. Nine papers are included in this year's proceedings.

Focused Session 2 was dedicated to design, modeling, simulation and characterization of ceramics and composites. 28 technical papers were presented regarding the prediction of crystal structure and phase stability; the characterization of interfaces and grain boundaries at the atomic scale; optimization of electrical, optical and mechanical properties; modeling of defects and related properties; design of materials and components at different length scales; as well as application of novel computational methods for processing. Three papers from this particular focused session are included in this issue.

Symposium 10 discussed new materials and the associated technologies related to thermal management. This included innovations in ceramic or carbon based materials tailored for either high conductivity applications (e.g., graphite foams) or insulation (e.g., ceramic aerogels); heat transfer nanofluids; thermal energy storage devices; phase change materials; and a slew of technologies that are required for system integration. Five papers were submitted for inclusion in this proceedings issue.

Symposium 11 addressed advances related to sensor technologies based on ceramic materials. The research scope ranged from advanced materials development, micro-fabrication technologies, sensor design, sensors and sensor networks, to packaging. The symposium included researchers from disciplines such as physics, chemistry, materials science, electronics, and instrumentation; as well as those who use sensors in energy, bio, space, buildings and etc, to highlight the latest developments and future challenges in this exciting interdisciplinary research field. Three papers were submitted and accepted for publication.

The editors wish to thank the symposium organizers for their time and labor, the authors and presenters for their contributions; and the reviewers for their valuable comments and suggestions. In addition, acknowledgments are due to the officers of the Engineering Ceramics Division of The American Ceramic Society and the 2011 ICACC program chair, Dr. Dileep Singh, for their support. It is the hope that this volume becomes a useful resource for academic, governmental, and industrial efforts.

WALTRAUD M. KRIVEN, *University of Illinois at Urbana-Champaign, Urbana, Illinois, USA*

ANDREW GYEKENYESI, *NASA Glenn Research Center, Cleveland, Ohio, USA*

JINGYANG WANG, *Institute of Metal Research, Chinese Academy of Sciences, CHINA*

Introduction

This CESP issue represents papers that were submitted and approved for the proceedings of the 35th International Conference on Advanced Ceramics and Composites (ICACC), held January 23–28, 2011 in Daytona Beach, Florida. ICACC is the most prominent international meeting in the area of advanced structural, functional, and nanoscopic ceramics, composites, and other emerging ceramic materials and technologies. This prestigious conference has been organized by The American Ceramic Society's (ACerS) Engineering Ceramics Division (ECD) since 1977.

The conference was organized into the following symposia and focused sessions:

Symposium 1	Mechanical Behavior and Performance of Ceramics and Composites
Symposium 2	Advanced Ceramic Coatings for Structural, Environmental, and Functional Applications
Symposium 3	8th International Symposium on Solid Oxide Fuel Cells (SOFC): Materials, Science, and Technology
Symposium 4	Armor Ceramics
Symposium 5	Next Generation Bioceramics
Symposium 6	International Symposium on Ceramics for Electric Energy Generation, Storage, and Distribution
Symposium 7	5th International Symposium on Nanostructured Materials and Nanocomposites: Development and Applications
Symposium 8	5th International Symposium on Advanced Processing & Manufacturing Technologies (APMT) for Structural & Multifunctional Materials and Systems

Symposium 9	Porous Ceramics: Novel Developments and Applications
Symposium 10	Thermal Management Materials and Technologies
Symposium 11	Advanced Sensor Technology, Developments and Applications
Symposium 12	Materials for Extreme Environments: Ultrahigh Temperature Ceramics (UHTCs) and Nanolaminated Ternary Carbides and Nitrides (MAX Phases)
Symposium 13	Advanced Ceramics and Composites for Nuclear and Fusion Applications
Symposium 14	Advanced Materials and Technologies for Rechargeable Batteries
Focused Session 1	Geopolymers and other Inorganic Polymers
Focused Session 2	Computational Design, Modeling, Simulation and Characterization of Ceramics and Composites
Special Session	Pacific Rim Engineering Ceramics Summit

The conference proceedings are published into 9 issues of the 2011 Ceramic Engineering & Science Proceedings (CESP); Volume 32, Issues 2-10, 2011 as outlined below:

- Mechanical Properties and Performance of Engineering Ceramics and Composites VI, CESP Volume 32, Issue 2 (includes papers from Symposium 1)
- Advanced Ceramic Coatings and Materials for Extreme Environments, Volume 32, Issue 3 (includes papers from Symposia 2 and 12)
- Advances in Solid Oxide Fuel Cells VII, CESP Volume 32, Issue 4 (includes papers from Symposium 3)
- Advances in Ceramic Armor VII, CESP Volume 32, Issue 5 (includes papers from Symposium 4)
- Advances in Bioceramics and Porous Ceramics IV, CESP Volume 32, Issue 6 (includes papers from Symposia 5 and 9)
- Nanostructured Materials and Nanotechnology V, CESP Volume 32, Issue 7 (includes papers from Symposium 7)
- Advanced Processing and Manufacturing Technologies for Structural and Multifunctional Materials V, CESP Volume 32, Issue 8 (includes papers from Symposium 8)
- Ceramic Materials for Energy Applications, CESP Volume 32, Issue 9 (includes papers from Symposia 6, 13, and 14)
- Developments in Strategic Materials and Computational Design II, CESP Volume 32, Issue 10 (includes papers from Symposium 10 and 11 and from Focused Sessions 1, and 2)

The organization of the Daytona Beach meeting and the publication of these proceedings were possible thanks to the professional staff of ACerS and the tireless dedication of many ECD members. We would especially like to express our sincere thanks to the symposia organizers, session chairs, presenters and conference attendees, for their efforts and enthusiastic participation in the vibrant and cutting-edge conference.

ACerS and the ECD invite you to attend the 36th International Conference on Advanced Ceramics and Composites (http://www.ceramics.org/daytona2012) January 22–27, 2012 in Daytona Beach, Florida.

SUJANTO WIDJAJA AND DILEEP SINGH
Volume Editors .
June 2011

Geopolymers and Other Inorganic Polymers

EFFECT OF EXTERNAL AND INTERNAL CALCIUM IN FLY ASH ON GEOPOLYMER FORMATION

Kiatsuda Somna
Faculty of Engineering and Architecture, Rajamangala University of Technology Isan, Nakornratchasima, Thailand, 30000

Walairat Bumrongjaroen
Vitreous State Laboratory, The Catholic University of America, Washington, DC, USA, 20064

ABSTRACT

Calcium in fly ash can be present as a glass component or as a free lime. The presence of calcium in fly ash glass so called "internal calcium" can have effect on the leaching of the glass and consequently the availability of Si and Al for geopolymer product formation. Whereas, the presence of calcium as a free lime in fly ash so called "external calcium" would have indirect effect on leaching as this fly ash contains higher Si and Al contents. In addition to the effect of calcium on the leaching, the source of calcium can have effect on the product formation. To study these two effects, the synthetic fly ash glasses with low and high calcium content were prepared to be used for leaching test and for formulating geopolymer. The specimen with external calcium was prepared by adding CaO to the geopolymer made with synthetic fly ash with low calcium. The specimen with internal calcium was prepared by using synthetic fly ash with high calcium content in the geopolymer mix. Synthetic fly ash was used instead of real fly ash to control the glass composition in fly ash and to avoid the effect of calcium from mineral phases. Synthetic fly ash glasses were prepared from lab grade chemical oxide as a starting material for geopolymer matrices. The mole ratio of Si:Al was kept at 1.84:1. The fly ash glass was ground and then mixed with NaOH solution. The products were characterized by FTIR technique to study the polymerization of the gel. The presence of calcium composite gel was determined by the difference in FTIR pattern before and after acid leaching which is specific for calcium composite gel. The compressive strength of geopolymer was investigated to study the contribution of these gels to strength. The percentage of C-S-H gel and N-A-S-H gel formed were determined and correlated with the compressive strength of geopolymer. It was found that the type of glass and the source of calcium have effect on geopolymer product formation. The geopolymer made from low calcium fly ash and external calcium has higher compressive strength than the one made from high calcium fly ash. This can be due to the leaching of the glass as the leaching result shows that low calcium glass can provide more Si and Al content for N-A-S-H and other zeolite formation. As for the effect of the source of calcium on the product formation, the FTIR analysis shows that the internal calcium in the high calcium glass forms more CSH gel than the external calcium in the low calcium glass. This could be because the external calcium might precipitate in high pH solution while the internal calcium does not.

INTRODUCTION

Geopolymer or inorganic polymer is aluminosilicate framework. The term of geopolymer was first applied by Davidovids, 2008[1]. Reaction of geopolymer is called geopolymerization which involves a chemical reaction in a highly alkali such as sodium hydroxide, potassium hydroxide and/or sodium silicate, potassium silicate solution with aluminosilicate reactive material such as metakaolin, fly ash. Under strong alkali solution, aluminosilicate materials are rapidly dissolved to form free tetrahedral of Si and Al (SiO_4 and AlO_4^-). In geopolymerization, water gradually splits out and tetrahedral of Si and Al are linked alternatively to yield polymeric precursors by sharing all oxygen atoms between two tetrahedral units[1]. In term of geopolymer, it was proposed that geopolymer is formed by geopolymerization of individual aluminate and silicate spicies[1]. The major difference between geopolymer and Portland cement is the presence of calcium in their structure. Calcium silicate hydrate (CSH) is the major product of Portland cement while calcium is not essential in any geopolymerization product. Many starting material of geopolymer composed of calcium such as Class C fly ash, and furnace slag.

Class F fly ash is a good candidate for starting material of geopolymer as it can provide alumina as well as silica to the geopolymer process. It is composed of 60-80% amorphous phase and some crystalline phase such as quartz, mullite, hematite and maghetite[2-4]. Class C fly ash as it contains higher calcium content can provide calcium in addition to silica and alumina to the matrix. The dissolved calcium can participate in three processes. First, it can be hold in the matrix inside the geopolymer product[5]. This calcium can interfere with geopolymer process and have effect on the property. Second, under strong alkali solution of geopolymer mix, the calcium hydroxide would precipitate as it has low solubility in water at high pH. Third, the remaining of calcium can form CSH gel when the alkalination induces the cleavage of siloxane link (Si-O-Si) that react with $Ca(OH)_2$ to form Ca-Di-siloxonate tobermorite or CSH gel[1,6-7]. Some researchers found that, calcium in fly ash improves mechanical property of geopolymer. It accelerates the hardening process and increases the strength[8-9]. Some groups found that adding $Ca(OH)_2$ to geopolymer mixture can improve mechanical property[6,10]. Notice that this property is the combined property of Ca-geopolymer, CSH gel and geopolymer gel. No work has been done to distinguish the ratio of these products. The source of calcium might have effect on the formation of these products. External calcium is the calcium that is added in to the mix while internal calcium is the intrinsic calcium in the fly ash glass. It is believed that the external calcium is more readily available than the internal calcium as the internal calcium inside the glassy phases might not dissolve all out at early age. Thus, the system with external calcium would have more calcium products than the system with internal calcium. Alonso, Palomo, Yip and Kumar found that both internal and external calcium can produce CSH gel, CAH and CASH gel[9,11-12].

In this research, the effect of fly ash glass type and the source of calcium on geopolymer production formation were studied. The fly ash with external calcium was prepared by synthesizing fly ash glass with low calcium and adding $Ca(OH)_2$ to obtain the same $Ca(OH)_2$ content as fly ash with high calcium, so called, internal calcium. The fly ash glass was synthesized from the chemicals in order to control the chemical composition. The products of geopolymer were characterized by FTIR, XRD and salicylic acid with methanol (SAM solution) leaching test. The compressive strength of geopolymer was also measured and related with other properties.

TECHNICAL APPROACH

Synthesizing fly ash glass

The fly ash glass was batched by lab grade chemical oxide. The glasses with low and high calcium were named SFALC and SFAHC, respectively. The mixture was melted at 1550°C in electric furnace. Fly ash glass was ground and sieved under 45 μm. The chemical composition of fly ash glass was determined by XRF technique. The mineralogy of glass was characterized by XRD technique. The molecular structure of fly ash glass was characterized by FTIR technique.

Reactivity Test

The glass samples were hydrated in 14 M NaOH to determine the reactivity. Initially about 1 g of glass was placed in contact with 100 ml of a 35 °C alkaline solution, leading to a glass surface area to solution volume (S/V) ratio at the beginning of the test of 100 cm^{-1}. The glass powder was hydrated up to 7 days at 35 °C. At the end of the leaching period the samples was filtered and then dried at room temperature before measuring the weight difference. The remaining fly ash was determined the chemical composition by XRF technique.

Synthesizing geopolymer paste

The geopolymer paste with internal calcium was prepared by mixing SFAHC with 14 M NaOH solution while that of external fly ash was prepared from the mix of SFALC with $Ca(OH)_2$ and 14 M NaOH solution. The mole ratio of Si:Al was 1.84:1. Calcium hydroxide 13.2 wt% was added into SFALC mixture to maintain the same molar content of Ca^{2+} in both mixtures. Both were continuously stirred for 5 minutes and, casted in cylindrical mold (30 mm in diameter and 60 mm in height). They were cured at 60°C for 2 hours. The hardened paste was demolded and kept at 36°C until the test age. The compressive strength of these specimens was tested at the age of 3, 7 and 28 days. The tested samples were ground and characterized by FTIR, XRD and SAM solution leaching.

Salicylic acid with methanol (SAM solution) leaching test

This test was first introduced by Takashima who used this technique to dissolve alite and belite in Portland cement. The solution of salicylic acid with methanol was prepared from 7 grams of salicylic acid and 40 ml of methanol. It was used to dissolve calcium bearing phase in geopolymer paste and report as the percentage of mass dissolved[13].

This leaching test was modified to study product of geopolymer pastes made from fly ash with high and low calcium content (GHC-internal and GLC-external in shorthand). The geopolymer pastes were ground and stirred for 3 hours. Then, it was filtered and dried at 100°C. Part of insoluble residue of geopolymer pastes after SAM solution leaching was characterized by XRF, FTIR and XRD to determine the product formed in geopolymer pastes. The soluble part was used to study the percentage of mass dissolved including calcium compound and new phases of GHC-internal and GLC-external by weight loss.

FTIR technique

Fourier transform infrared spectroscopy was chosen to study the gel products of geopolymer. FTIR was obtained from PerkinElmer instrument. The specimen was prepared by 0.001 mg of sample in 0.04 mg of KBr. Spectra analysis was performed over the range 4,000–400 cm^{-1} at the resolution of 4 cm^{-1}.

Steps to determine the product of geopolymer are as follows.
1. Synthetic fly ash glass was characterized by FTIR.
2. GHC-internal and GLC-external were characterized by FTIR and subtracted with FTIR spectra of fly ash glass. The possible products found in resultant FTIR spectra are calcium silicate hydrate gel (CSH gel), calcium aluminate silicate hydrate gel (CASH gel), geopolymer gel (NASH gel) and Zeolite as shown in Eq.(1)

$$\text{Geopolymer Pastes} - \text{Unreacted Fly Ash Glass} = \text{CSH gel} + \text{CASH gel} + \text{NASH gel} + \text{Zeolite} \qquad (1)$$

3. Insoluble residue of GHC-internal and GLC-external after SAM solution leaching was characterized by FTIR. They were subtracted with FTIR spectra of their fly ash glass. The possible products found from FTIR spectra are NASH gel and Zeolite as shown in Eq. (2)

$$\text{Insoluble Residue after SAM Sol}^n - \text{Unreacted Fly Ash Glass} = \text{NASH gel} + \text{Zeoliite} \qquad (2)$$

4. FTIR spectra of CSH gel and CASH gel was obtained by subtracting FTIR spectra obtained from Eq. (1) and Eq. (2) as shown Eq. (3).

$$\text{CSH gel} + \text{CASH gel} = (\text{CSH gel} + \text{CASH gel} + \text{NASH gel} + \text{Zeolite}) - (\text{NASH gel} + \text{Zeolite}) \qquad (3)$$

XRD technique

The XRD pattern was obtained using Thermo-Fisher X-TRA with CuK radiation. Specimens were scanned at 2θ from $5°$ to $80°$ as $0.02°$ step side at the rate of 2 sec per step. This data was used to provide fundamental information of GHC-internal, GLC-external and insoluble residue of GHC-internal and GLC-external after SAM solution leaching at the age of 28 days.

RESULT AND DISCUSSION

Properties of Glass

Fly ash glass was synthesized in order to control the glass composition in fly ash, particle size distribution and avoid inert particle of mineral phases. The chemical composition of glass was analyzed by XRF as shown in Table 1. The high calcium Class F fly ash (SFAHC) has $SiO_2 + Al_2O_3 + Fe_2O_3$ of 82.72% and CaO of 14.35%. The low calcium Class F fly ash (SFALC) has $SiO_2 + Al_2O_3 + Fe_2O_3$ of 92.55% and CaO of 4.04%. They are considered to be Class F fly ash as described by ASTM C618[14]. Particle size distributions of synthetic fly ashes (SFAHC and SFALC) are shown in Fig.1. SFAHC and SFALC were ground and sieved to pass 45 um sieve. Geopolymer prepared from fine fly ash and strong alkali solution is expected to have high performance. Fig. 2 presents XRD pattern of SFALC and SFAHC, respectively. There are amorphous silica peaks around 20 to 30° at 2θ. There is no evidence of crystalline phase. FTIR spectra of SFAHC and SFALC glasses are shown in Fig. 3. The major broad peak of SFALC at 1074 cm^{-1} is associated with Si-O asymmetric stretching vibration Q_3 units. The main spectra peak of SFAHC shifts to lower wavenumber at around 970 cm^{-1}, typical of Si-O Q_2 units[15].

Table 1 The chemical composition of low and high calcium fly ash glass

ID	Al_2O_3	CaO	Fe_2O_3	K_2O	MgO	Na_2O	SiO_2
SFAHC	22.98	14.35	12.30	1.65	0.626	0.544	47.44
SFALC	25.87	4.04	13.70	1.84	0.70	0.732	52.98

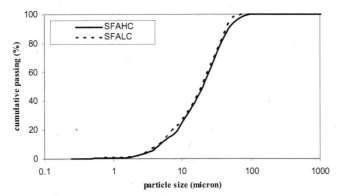

Figure 1 Particle size distribution of SFALC and SFAHC

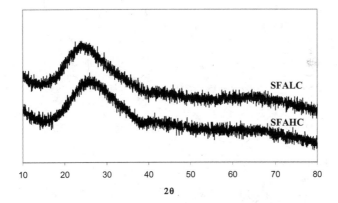

Figure 2 XRD patterns of SFAHC and SFALC

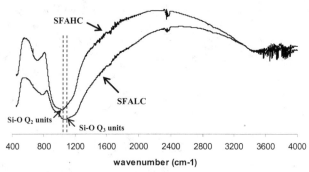

Figure 3 FTIR of synthetic fly ash glasses; SFAHC and SFALC

Reactivity of Fly Ash

The chemical analysis of the hydrated fly ash glass is shown in Table 2 below. The result shows that there is high percentage of Na in the solid part which was not present in the initial fly ash glass. This suggests that the sodium product was formed. The XRD patterns of both hydrated glasses indicate the presence of Na_2CO_3. Since sodium was not present in initial glass, all of the sodium content in the composition of hydrated glass can be removed. By subtracting the percentage of Na_2O out, the remaining is the composition of the hydrated glass. As Si is the main network component of the glass, the percentage of Si removal from glass is used as a measure of fly ash reactivity. The result is shown in Table 2. SFAHC glass is 14.4% reactive while SFALC glass is 28.7 % reactive.

Table 2 Normalized composition of SFAHC and SFALC without Na and the percentage of Si removal

ID	Al_2O_3	CaO	Fe_2O_3	K_2O	MgO	Na_2O	SiO_2	%Si removal
SFAHC-N	16.57	20.80	19.63	1.72	0.86	0	40.42	14.4
SFALC-N	18.38	8.00	31.58	2.69	1.58	0	37.77	28.7

FTIR analysis

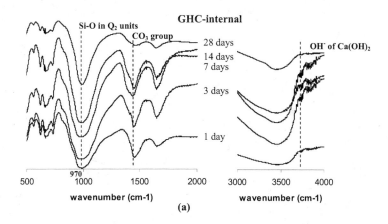

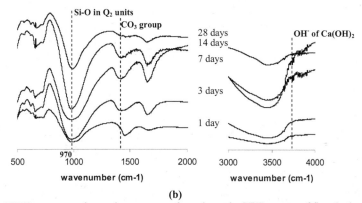

(b)

Figure 4 FTIR spectrum of geopolymer pastes after subtract by FTIR spectra of fly ash glass
(a) GHC-internal and (b) GLC-external

FTIR spectra of GHC-internal and GLC-external with time are shown in Fig. 5 (a) and Fig. 5 (b), respectively. In this research, 14.4% of SFAHC and 28.7% of SFALC are reactive. Thus 85.6% of SFAHC and 71.3% of SFALC in fly ash spectra intensity was used to subtract to geopolymer paste spectra in order to remove the unreacted fly ash glass spectra from the geopolymer spectra. After subtracted by FTIR of unreacted fly ash glass, the remaining FTIR spectra is the spectra of geopolymer products i.e. CSH gel, CASH gel, NASH gel and Zeolite.

The spectrum shows a main narrow band at around 966 cm^{-1} at the early age (1 day) as a result of Al effect which changes Si-O-Si into Si-O-Al linkage. It causes the broadening of the band around 970 cm[-1] [16]. This peak became somewhat appear sharper after the curing age increased due to there was an increasing Si leached out of fly ash, typical of the Si-O asymmetric stretching vibration generated by Q_2 units[15]. There is also sharp peak at 3640 cm[-1] in GLC-external, corresponding to $Ca(OH)_2$, which came from both calcium in CSH product and crystalline $Ca(OH)_2$. So CH peak cannot be used to support the formation of the CSH gel. The narrow band is also visible at around 1400 – 1450 cm[-1], typical of CO_3 group is attributed to $CaCO_3$ or Na_2CO_3 from reaction between $Ca(OH)_2$ or Na_2O and CO_2 in the atmosphere[7].

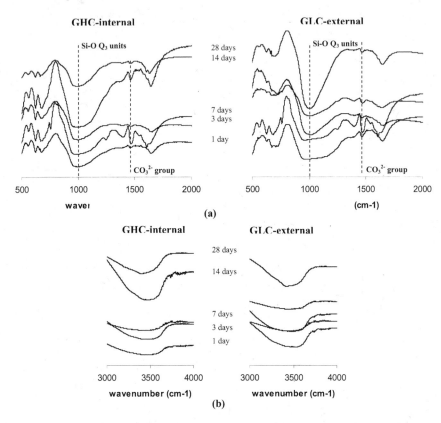

Figure 5 FTIR spectra of insoluble residue of geopolymer after SAM solution leaching and subtracted by FTIR spectra of fly ash glass
(a) GHC-internal and GLC-external at low wavenumber (400 – 2000 cm[-1])
(b) GHC-internal and GLC-external at high wavenumber (3000 – 4000 cm[-1])

Fig. 5 (a) and Fig. 5 (b) show FTIR spectra of insoluble residue of GHC-internal and GLC-external after SAM solution leaching and subtracted by FTIR spectra of 85.6% and 71.3% of fly ash glass at low wavenumber (400 – 2000 cm[-1]) and high wavenumber (3000-4000 cm[-1]), respectively.

In our work, it was found that SAM solution can dissolve Ca-compound such as CSH gel, CASH gel, $Ca(OH)_2$ and some Na compound such as remaining NaOH and Na_2CO_3. This procedure is based on the formation of a soluble calcium complex with Ca in CSH gel CASH gel and $Ca(OH)_2$. $CaCO_3$ cannot be dissolved by SAM solution due to its less solubility in water. It is extremely stable as a solid and water doesn't have sufficient solvating capability to cause the ions to separate and come into solution. In addition, both NASH gel and Zeolite is complex structures, SAM solution cannot attack their structure. Consequently, the resultant FTIR spectra of insoluble residue of geopolymer after SAM solution leaching and subtracted by FTIR spectra of fly ash glass could be the spectra of NASH gel and Zeolite.

There is a broad band appearing in high wavenumber at around 1000 cm^{-1}, indicating Si-O-Si stretching vibrations of geopolymerization[17]. It is possible be NASH gel occurring in the geopolymer. The intensity of the major peak (around 1000 cm^{-1}) slightly increased as the curing age increased. At 28 days, GLC-external has higher intensity at 1000 cm^{-1} than GHC-internal due to high Si and Al content in SFALC. As such, the GLC-external could produce more NASH gel than GHC-internal. There is the band characteristic of carbonate group at around 1450 cm^{-1}. It was supposed to be $CaCO_3$ from reaction between $Ca(OH)_2$ and CO_2 in atmosphere which cannot be dissolve by SAM solution due to low solubility property.

In addition, Fig.6 presents XRD pattern of GHC-internal, GLC-external and insoluble residue of GHC-internal and GLC-external after SAM solution leaching at the age of 28 days. Crystalline phase of Zeolite was found in both XRD pattern of geopolymer pastes and XRD pattern of insoluble residue geopolymer paste after SAM solution leaching in form $Na_2O.Al_2O_3.1.7SiO_2.1.7H_2O$. Zeolite is alumino-silicate mineral which possibly came from the amorphous geopolymeric products when the reaction condition is suitable[18]. It should be noted that SAM solution can not dissolve NASH gel and Zeolite due to both the structure of NASH gel and Zeolite have Si-O in Q_3 units (tetrahedral).

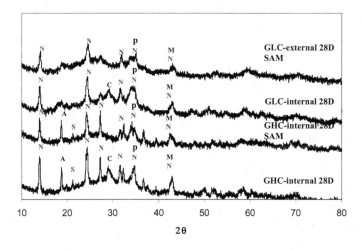

N = Sodalite S = Laumontite P = Portlandite M = Periclase
C = Sodium Silicate A = Aluminum Hydroxide

Figure 6 XRD pattern of GHC-internal, GLC-external and insoluble residue of GHC-internal and GLC-external after SAM solution leaching

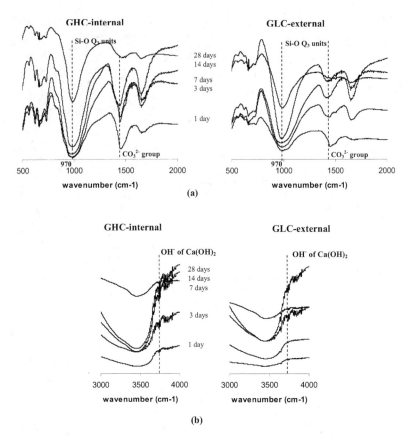

Figures 7 FTIR spectra resulting from subtracting the spectra in Figure 5 from Figure 4
(a) GHC-internal and GLC-external at low wavenumber (400 – 2000 cm⁻¹)
(b) GHC-internal and GLC-external at high wavenumber (3000 – 4000 cm⁻¹)

Fig. 7 shows the FTIR spectra resulting from subtracting the spectra insoluble residue (Figure 5) from geopolymer (Figure 4). The resultant FTIR spectra in this part are the spectra of CSH gel and CASH gel.

The major peak of this part is shift to lower wavenumber at around 970 cm⁻¹, corresponding to the Si-O asymmetric stretching bands in Q_2 units, typical of CSH gel[15]. Both GHC-internal and GLC-external have peak of at 970 cm⁻¹ at the early age to 28 days. At the early age, GHC-internal has higher intensity at 970 cm⁻¹ than GLC-external. This could be because SFAHC released Si and Ca more rapidly to form CSH gel at the early age while SFALC released Si slowly to react with the external calcium that was readily available outside the glass. Some external calcium might precipitate as evidenced from $Ca(OH)_2$ band in its FTIR spectra. At the age of 28 days, GHC-internal has higher intensity at 970 cm^{-1} than GLC-external because high Ca content in glass could produce more CSH gel or CASH gel than GLC-external sample that Ca was added in the mix. It was also found peak at around 3640 cm^{-1} which attributed to $Ca(OH)_2$ from both CSH gel and crystalline $Ca(OH)_2$. It supports the presence of CSH gel in both samples. In addition, the peak at 1450 cm^{-1} corresponds to carbonate group (CO_3) which could be Na_2CO_3. This sodium salt can be dissolved by SAM as it has weak ionic bond and high solubility in water. SAM solution cannot dissolve $CaCO_3$ as it has stronger bond and is more stable as solid phase. Thus, SAM solution can dissolve calcium compound including CSH gel, CASH gel and $Ca(OH)_2$ and sodium compounds including sodium silicate and Na_2CO_3.

In summary, the products of GHC-internal and GLC-external can be determined by FTIR spectra. NASH gel and Zeolite were found by FTIR spectra of insoluble residue of geopolymer after SAM solution. By measuring the intensity of major peak at around 1000 cm^{-1} which is the sign of NASH gel and Zeolite, GLC-external has higher NASH gel than GHC-internal. This can be because SFALC has high Si and Al content for the formation of NASH and Zeolite. The CSH gel and CASH gel can be determined from FTIR spectra of the part of gel dissolve by SAM solution. The relative amount of CSH and CASH gel is determined from the intensity of major peak at around 970 cm^{-1}. It was found that GHC-internal has higher CSH gel or CASH gel than GLC-external due to the high calcium content in SFAHC.

Analysis of C-S-H, C-A-S-H gel by SAM solution

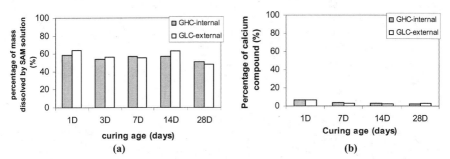

Figure 8 Percentage of mass dissolved by SAM solution (a) Percentage of mass dissolved, (b) Percentage of calcium compound

The percentage of mass dissolved by SAM solution of GHC-internal and GLC-external geopolymer are shown in Fig. 8(a). The percentage of mass dissolved in both GHC-internal and GLC-external is around 50-60%. The analysis of mass dissolved by SAM solution was intended to determine the CSH and CASH gel content in the geopolymer pastes. To determine the major elements in the product dissolved from the matrix, the soluble part was analyzed by XRF. High percentage of Na_2O around 45 to 60% was found in soluble part. Thus the percentage of mass dissolved by SAM solution cannot be used to determine CSH and CASH gel. To confirm that NASH gel cannot be dissolved by SAM solution, the synthetic NASH gel was prepared by sol-gel method by mixing 1 M of $Al((NO)_3)_3.9H_2O$ for aluminum with 1 M of sodium metasilicate for sodium and silica. Aluminum nitrate was dissolved in water and slowly added into sodium silicate solution. NASH gel was characterized by FTIR and measured the mass loss after SAM solution leaching. It was found that NASH gel and Zeolite cannot be dissolved by SAM solution. Insoluble residue of NASH gel after

SAM solution leaching was also characterized by FTIR and found both FTIR spectra of NASH gel and insoluble residue of NASH gel after SAM solution present Si-O Q_3 units, typical geopolymer or Zeolite form. Thus, the Na compound that was dissolved out can be the remaining NaOH and Na_2CO_3 or the new Na phase formed in the reaction. It also shows that not all Na participates in geopolymerization of NASH gel. The percentage of Na was subtracted by percentage of mass dissolved to obtain the percentage of Ca compound as shown in Fig. 9(b). There are small amount of Ca compound dissolved. It was found that the dissolved calcium from GHC-internal is higher than that of GLC-external. It cannot be concluded from this result that GHC-internal has higher CSH gel as the $Ca(OH)_2$ can be dissolved by SAM solution also.

Compressive Strength

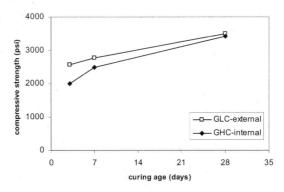

Figure 9 Compressive strength of GHC-internal and GLC-external

Fig. 9 shows the compressive strength of GHC-internal and GLC-external. The data of compressive strength was correlated with intensity of FTIR spectra products of GHC-internal and GLC-external as shown in Fig. 5 and Fig.7. The compressive strength of GLC-external is higher than that of GHC-internal. According to FTIR of GLC-external, it has higher intensity of NASH product than GHC-internal while GHC-internal has higher intensity of CSH gel. It could be because this Class F fly ash with low calcium (SFALC) released more silica and alumina to produce NASH gel than that of GHC-internal as shown in FTIR spectra (Fig.5 (a)). Theoretically, this SFALC fly ash should dissolve Si, Al slowly as this glass contains high Si and Al which are network former of the glass. However, it is evidently shown that the geopolymer made from SFALC fly ash has higher strength and higher intensity of NASH gel. This shows that it can release more Si and Al for geopolymer reaction. The additional strength at the early age could be the contribution from CSH gel which was formed from the reaction between dissolved silica and external calcium. Its compressive strength then increased at later age as the geopolymer reaction continued to proceed. The compressive strength of GHC-internal was lower than that of GLC-external at all ages although it has higher intensity of CSH gel and CASH gel. This can be because it has lower Si and Al content to form NASH gel. GHC-internal which has calcium in fly ash glass can form CSH gel and NASH gel, simultaneously as evidenced by CSH gel and NASH gel in FTIR spectra at early age.

SUMMARY

FTIR was used together with XRD and SAM leaching to identify the product of geopolymer i.e. CSH gel, CASH gel, NASH gel and Zeolite. Both of GHC-internal and GLC-external have CSH gel CASH gel, NASH gel and Zeolite. Both glass type and calcium source have the effect on geopolymer product formation. The low calcium fly ash (GLC-external) has higher Si removal percentage in alkali solution than the high calcium fly ash (GHC-internal). This dissolved silica can react with Na or Ca that are available in the solution to form NASH or CSH gel respectively. The geopolymer made of GLC-external has higher strength than the one made of GHC-internal. This is because its glass has higher silica and alumina that can release out to form NASH gel. It is not because of CSH gel as the FTIR analysis shows that GLC-external system has lower intensity of CSH gel than GHC-internal even though it has calcium added to the system. The added calcium might precipitate in this high pH solution and does not form CSH gel. In GHC-internal sample, its internal calcium was more readily available to react with silica. Thus, the dissolved silica was consumed by this calcium to form CSH gel. This indicates that the presence or location of calcium designates the formation of CSH gel in geopolymer. In conclusion, both glass components and the source of calcium has an effect on the formation of geopolymer products in which the low calcium (high Si and Al) glass has higher Si removal percentage than the high calcium (low Si and Al) glass and the internal calcium can react with dissolved Si to form more C-S-H gel than external calcium.

ACKNOWLEDGEMENTS

The authors would like to thank Vitreous State Laboratory, Catholic University of America, Washington, DC, USA for giving financial support to do this research and Chemistry Department, the Catholic University of America for giving access to FTIR instrument.

REFERENCES

[1] J. Davidovit, Geopolymer chemistry and application, Institute Geopolymer, 16 rue Galilee F-02100 Saint-Quentin, France, 585 pages (2008).

[2] C. Shi, R. L. Day, Acceleration of the reactivity of fly ash by chemical activation, *Cement and Concrete Research*, 25 (1), 15 -21 (1995).

[3] M. Ilica, C. Cheeseman, C. Sollarsb and J. Knightb, Mineralogy and microstructure of sintered lignite coal fly ash, *Fuel*, 82(3), 331-336 (2003).

[4] A. Ferna´ndez-Jime´nez, A. G. Torre, A. Palomo, G. Lo´pez-Olmo, M. M. Alonso, M. A. G. Aranda, Quantitative determination of phases in the alkaline activation of fly ash. Part II: Degree of reaction, *Fuel,* 85, 1960–1969 (2006).

[5] D. Hardjito, B.V. Rangan, Development and properties of low-calcium fly ash-Based geopolymer concrete, Research Report GC1, Faculty of engineering, Curtin University of Technology, Perth, Australia, 90 pages (2005).

[6] A. Buchwald, K. Dombrowski, M. Weil, The influence of calcium content on the performance of geopolymeric binder especially the resistance againt acid. 4th international conference on geopolymers, 29.6.-1.07.05, sy, Quentin, France.

[7] K. Komnitsas, D. Zaharaki, Geopolymerisation: A review and prospects for the minerals industry, *Minerals Engineering* 20, 1261 – 1277 (2007).

[8]H. Xu, J.S.J. Van Deventer,Geopolymerisation of multiple minerals, *Minerals Engineering*, 15, 1131 – 1139 (2000a).

[9]C. K. Yip, G. C. Lukey, J. S. J. Van Deventer, The coexistence of geopolymeric gel and calcium silicate hydrate at the early stage of alkaline activation, *Cement and Concrete Research,* 35, 1688– 1697 (2005).

[10]J. Temuujin, A. Van Riessen, R. Williams, Influence of calcium compounds on the mechanical properties of fly ash geopolymer pastes, *Journal of Hazardous Materials,* 167, 82–88 (2009).

[11]S. Alonso and A. Palomo, Alkaline activation of metakaolin and calcium hydroxide mixtures: influence of temperature, activator concentration and solids ratio, *Materials Letters*, 47 (1-2), 55-62 (2001).

[12]S. Kumar, R. Kumar, S. P. Mehrotra, Influence of granulated blast furnace slag on the reaction, structure and properties of fly ash based geopolymer, *Journal of Material Science*, 45, 607–615 (2010).

[13]W.A. Gutteridge, On the dissolution of the interstitial phases in Portland cement, *Cement and Concrete Research*, 9, 319 – 324 (1979).

[14] ASTM. (2001). "Standard specification for coal fly ash and raw or calcined natural pozzolan for use as a mineral admixture in concrete." ASTM C 618, Philadelphia.

[15]I. G. Loderio, D. E. Macphee, A. Palomo, A. Fernandez-jimenez, Effect of alkalis on fresh C-S-H gels. FTIR analysis, *Cem and Concr Res*, 39, 147 – 153 (2009).

[16]I. G. Loderio, A. Fernandez-jimenez, A. Palomo, D.E. Macphee, Effect on fresh C-S-H gels of the simultaneous addition of alkali and aluminum, *Cem and Concr Res*, 40, 27-32 (2010).

[17]U. Rattanasak, P. Chindaprasirt, Influence of NaOH solution on the synthesis of fly ash geopolymer, *Miner Eng*, 22, 1073 – 1078 (2009).

[18]Y. Zhang, W. Sun and Z. Li, Infrared spectroscopy study of structural nature of geopolymeric products, *Journal of Wuhan University of Technology--Materials Science Edition*, 23(4), 522-527 (2010).

SYNTHESIS AND THERMAL PROPERTIES OF FLY-ASH BASED GEOPOLYMER PASTES AND MORTARS

Ch. Panagiotopoulou, A. Asprogerakas, G. Kakali, S. Tsivilis
National Technical University of Athens, School of Chemical Engineering,
Zografou Campus, 15773 Athens, Greece

ABSTRACT

This work concerns the use of fly ash (coming from the power station at Megalopolis, Greece), as raw material for the synthesis of inorganic polymers and it is part of a research project concerning the exploitation of Greek minerals and by-products in geopolymer technology. Taguchi experimental designing model was applied in order to study the synergetic effect of selected synthesis parameters on the compressive strength development of fly ash based geopolymers. The experimental design involved the variation of three control factors in three levels. The selected factors and the corresponding level range were: i) the alkali to aluminum ratio in the starting mixture, $0.8 \leq R/Al \leq 1.2$, ii) the kind of alkali ion, $0 \leq Na/(Na+K) \leq 1.0$ and iii) the concentration of silicon in the activation solution, $1.0 \leq [Si]/R_2O \leq 2.0$. The above design procedure led to the conduction of 9 experiments. The compressive strength of geopolymers was measured and the final products were also examined by means of XRD. In addition, the thermal behavior of geopolymer pastes and mortars was evaluated on the basis of dimensional, mass and compressive strength changes after thermal treatment at temperature up to $800°C$.

As it is concluded, the optimal synthesis conditions were $R/Al=1.0$, $Na/(Na+K)=1.0$ and $[Si]/R_2O=1.0$, while the factor having the highest impact on the development of compressive strength was the $[Si]/R_2O$ ratio. The samples experienced significant strength loss after thermal treatment at $600°C$, while at higher temperature, the initially amorphous structure was replaced by nepheline and gehlenite

INTRODUCTION

The need for construction materials that possess improved fire-resisting properties led professor Joseph Davidovits to the synthesis of new materials which he named geopolymers[1]. Geopolymerisation concerns the reaction of an aluminosilicate mineral and a silicate solution in a highly alkaline environment. The synthesis and chemical composition of geopolymers are similar to those of zeolites, but their microstructure is amorphous to semi-crystalline[2]. Theoretically, any aluminosilicate material can undergo geopolymerisation under certain circumstances. Previous works have reported the formation of geopolymers from natural minerals [2-5], calcined clays [6,7], industrial by-products [8-11] or a combination of them [12-16].

The formation of geopolymers involves a chemical reaction between an aluminosilicate material and sodium silicate solution in a highly alkaline environment. The exact mechanism of this reaction is not yet fully understood, but it is believed to be a surface reaction consisting of four main stages: (1) the dissolution of solid reactants in an alkaline solution releasing Si and Al atoms, (2) the diffusion of the dissolved species through the solution, (3) the polycondensation of the Al and Si complexes with the added silicate solution and the formation of a gel and (4) the hardening of the gel that results to the final polymeric product. Stages (2) to (4) cannot be monitored since the procedures cannot be stopped and the products cannot be isolated. Any parameter that can affect any of the above mentioned stages can also affect the properties of the synthesized geopolymers. Many researchers studied the parameters affecting the synthesis and properties of the final products [5,17,18].

Factors, such as the curing conditions and the ratios of starting materials, strongly affect the structure and strength development of geopolymers [19-21]. Both the curing temperature and the curing time are considered to influence the properties of the produced geopolymers since the synthesis of geopolymers depends on the temperature of polycondensation and the time that the formatted gel remains in this temperature. In low temperatures amorphous or glassy structures are formed resulting in poor chemical properties. Temperatures from 35°C to 85°C lead to the formation of amorphous to semi-crystalline structures that possess good physical, thermal and mechanical properties while curing in temperatures higher than 100°C results to semi-crystalline structures that have excellent properties [22,23]. The effect of the concentration of silicon in the alkali activating solution has been investigated on systems based both on fly-ash [24] and metakaolin [6,25] as solid aluminosilicate sources. In addition, the kind of the alkali cation is also found to affect the reactivity of the raw material and the incorporation of Al in the geopolymeric matrix [26]. Usually, the effect of synthesis parameters is studied by changing one factor at a time. However geopolymerization is a complicated and dynamic process and the synthesis parameters seem to have a combinational effect. In this work the Taguchi experimental designing model was applied in order to study the synergetic effect of selected synthesis parameters on the strength development of fly ash based geopolymers.

Due to their ceramic-like properties, geopolymers are believed to possess good resistance to elevated temperatures. Published research in fly ash– based geopolymer in the area of fire resistance and exposure to elevated temperatures is of great interest and quality, but it is limited and there is a disagreement in many cases [27,28].

The objective of this work is, by using a geopolymer synthesis that possess good mechanical properties in room temperatures, to examine its behaviour to elevated temperatures as well as to assess the effect of the addition of sand on the thermal stability of the synthesised geopolymer- aggregate composites. The raw material used for the conduction of the experimental is fly ash. The development of fly ash based geopolymers can contribute to the exploitation of fly ash and its transformation into high added value product. Besides, it is believed that geopolymer binders can in general deliver 80% or greater reduction in CO_2 emission and consume 60% less energy, compared to OPC [17,29,30].

EXPERIMENTAL

Fly ash used for geopolymer synthesis comes from the power station at Megalopolis, Greece and its chemical composition is presented in Table I. This material consists mainly of quartz (SiO_2) and feldspars ($NaAlSi_3O_8$) while cristobalite, anhydrite, ghelenite and maghemite are found in smaller quantities. Fly ash was previously ground and its mean particle size (d_{50}) was approximately 22 μm. This is a typical fineness of fly ash when used in construction technology (as main constituent in blended cements).

Table I. Chemical composition of fly ash (% w/w)

SiO_2	Al_2O_3	Fe_2O_3	CaO	MgO	K_2O	Na_2O	SO_3	L.O.I.
54.18	15.92	7.94	7.08	6.08	0.94	2.26	0.64	1.93

The implementation of the Taguchi experimental designing model allows the investigation of the synergetic effect of the selected parameters, conducting the minimum number of experiments. The factors selected to be investigated were those determined by previous study[31] as the most influential on the development of the compressive strength of geopolymers, namely the molar ratios R/Al, Na/(Na+K) and the Si content of the activation solution [Si]/R_2O (R: Na or K). The experimental design involved their variation in three levels, while their variation extent was selected according to physical and chemical restrictions. The variation levels of the selected parameters are presented in Table II.

Table II. Investigated parameters and variation levels

Parameters	Level 1	Level 2	Level 3
Alkali to aluminum ratio (R/Al)	0.8	1.0	1.2
Alkali kind Na/(Na+K)	0.0	0.5	1.0
Content of soluble silica into sol gel [Si]/R_2O	1.0	1.5	2.0

The geopolymer samples were prepared using an aqueous activation solution containing sodium and/or potassium hydroxide and commercial water silica solution (50% w/w). The activation solutions were stored for a minimum of 24 h prior to use, to allow equilibrium. The raw material and the activation solution were mechanically mixed to form homogenous slurry which was transferred to cubic moulds and mildly vibrated. In all mixtures the ratio $m_{solids}/m_{liquids}$ was kept to 2.5. The specimens were left for 2 h at ambient temperature before they were cured at 70°C for 48 h. These curing conditions were found to be optimal in a previous work [32]. Their compressive strength was measured after 7 days and their structure was examined using XRD.

The second part of the experimental concerns the thermal stability of geopolymer pastes and geopolymer/aggregate composites. The synthesis conditions were those defined as optimal in the first part of this work. The sand used for the synthesis of the composites was calcareous sand of domestic origin. The specimens were subjected to temperatures of up to 800°C at a gradual incremental rate of approximately 3°C/min from room temperature. As soon as the target temperature was attained, the specimens were left for an additional 2 h before the furnace was shut down to allow the specimens in the furnace to cool down to room temperature. In the meantime, a set of the reference specimens were left undisturbed at room temperature for comparative study. After the thermal treatment, the mass and volume changes were recorded and the compressive strength of the specimens was measured. The ultrasonic pulse velocity test (apparatus: 58-E48, Controls Testing Equipments Ltd) was used as a measure of internal soundness of the samples. The measurements were carried out before and after each thermal treatment. Mass, volume, strength and ultrasonic pulse velocity measurement, presented in this work, are the average of three specimens.

The structural changes were investigated by means of XRD. X-ray powder diffraction patterns were obtained using a Siemens D-5000 diffractometer, CuK_{a1} radiation (λ= 1.5405Å), operating at 40kV, 30mA.

RESULTS AND DISCUSSION
Optimization of synthesis conditions

Table III shows the synthesis parameters as defined by Taguchi model and the compressive strength of the corresponding geopolymers.

Table III. Synthesis conditions and compressive strength of geopolymers

Experiments	R/Al	Na/(Na+K)	[Si]/R_2O	Compressive strength (MPa)
Experiment 1	0.8	0.0	1.0	34.0
Experiment 2	0.8	0.5	1.5	39.5
Experiment 3	0.8	1.0	2.0	37.4
Experiment 4	1.0	0.0	1.5	33.7
Experiment 5	1.0	0.5	2.0	32.8
Experiment 6	1.0	1.0	1.0	58.7
Experiment 7	1.2	0.0	2.0	28.0
Experiment 8	1.2	0.5	1.0	46.3
Experiment 9	1.2	1.0	1.5	45.1

The impact of each factor on the development of compressive strength was defined through the Taguchi mathematical processing and is presented in Table IV. The effect of each parameter on the development of compressive strength is presented in Figure 1.

Table IV: Contribution of the studied parameters to the development of the compressive strength

Parameter	R/Al	Na/(Na+K)	[Si]/R$_2$O
Contribution (%)	6.3	58.7	35.0

The parameter that seems to have the greatest impact on the development of the compressive strength is the kind of alkali ion. As it can be seen, Na-geopolymers possess better properties than those containing K or a mixture of K and Na. The second important parameter is the silicon content in the alkaline solution. The optimal value of this parameter is 1 and any further increase up to 2 induces a negative effect on the geopolymers' compressive strength. The R/Al ratio seems to have a marginal effect on the mechanical properties of the specimens. Nevertheless, the increase of the R/Al ratio from 0.8 to 1.0 improves the compressive strength of geopolymers. Further increase of the R/Al ratio has a negative effect on the development of the compressive strength. The optimum synthesis conditions for this type of fly ash, as it arises by the Taguchi mathematical processing and can be seen in Figure 1 is: R/Al=1, [Si]/R$_2$O=1 and Na/(Na+K)=1.

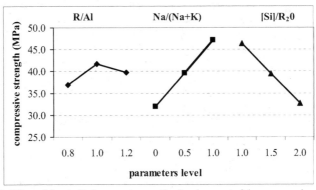

Figure 1. Effect of the studied parameters on the development of the compressive strength

Geopolymerization involves mainly two simultaneous reactions: the Si and Al dissolution from the raw material and the reorganization of these species into a three-dimensional amorphous network [33]. All parameters that affect these two reactions also affect the properties of the final product. The role of the alkalis during the synthesis is dual: they provoke the dissolution of Si and Al species from the raw material and they also participate in the formation of the aluminosilicate network playing a charge-balancing role. In the case of this type of fly ash, which has elevated calcium content, besides the Na and/or K species, Ca also moves into the solution, complicating the geopolymerization process. Calcium can react with the silicate or the aluminate species, participating in the formation of either the geopolymeric network or the C-S-H phases (as in the case of slag) [34]. As it is concluded by the study of the leaching behavior of fly ash, the increase of alkalinity does not induce an increase of the leaching ability of fly ash [32]. This is probably the reason that the R/Al ratio has only a limited effect on fly ash geopolymerisation.

The presence of sodium instead of potassium leads to the formation of geopolymers with higher compressive strength. The type of alkali ion is connected to the extent of dissolution as well as the polycondensation of the geopolymeric gel. The leaching behaviour of fly ash is not affected by the kind of alkali, therefore the observed effect of alkali ion is rather connected with the polycondensation process [32]. According to our experience, the effect of alkali ion is strongly related to the kind of fly ash and especially its chemical composition.

As it was concluded by previous studies, the presence of an adequate amount of initially dissolved silica is crucial for the development of the compressive strength of geopolymers. The absence or limited presence of dissolved silica seems to favor the appearance of zeolites within the geopolymers matrix while the presence of excess silica inhibits the dissolution of the raw material [19]. In this study, in order to avoid the appearance of zeolitic phases due to low silicon content, the [Si]/R_2O ratio varies from 1.0 to 2.0. The absence of zeolitic phases in all geopolymer samples can be confirmed by the XRD patterns. As it can be seen in Figure 1, increase of the initially dissolved silica leads to a decrease of their compressive strength. The presence of high amounts of initially dissolved silica leads to almost saturated solutions, inhibiting further dissolution of Si and Al from the raw material. In this case, the dissolution reaction of fly ash shifts to the left, leaving unreacted material and affecting the microstructure of the geopolymers as it was also confirmed in our previous studies [19].

The XRD patterns of geopolymers do not indicate significant differences, as far as their mineral composition is concerned. The XRD patterns of Na-geopolymers are presented in Figure 2. As it is seen, the increase of Na content in the activation solution (experiment 9) leads to the formation of natrite.

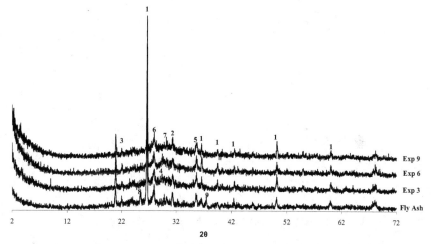

Figure 2: XRD patterns of Na-geopolymer pastes synthesised
(1: quartz, 2: gehlenite, 3: cristobalite, 4: calcite, 5: maghemite, 6: feldspars, 7: natrite, 8: anhydrite, 9: lime)

As it can be seen in Figure 2 the geopolymer matrix of all geopolymers contain only some of the crystalline phases of the raw material: quartz, feldspars, cristobalite, ghelenite and maghemite. The Ca-containing phases of lime and anhydrite of the fly ash pattern disappear in the case of geopolymers, accompanied by the appearance of the new phase of calcite. The peaks attributed to calcite are more obvious and better crystallined in the case of K-geopolymers, while in Na-geopolymers only traces of calcite can be detected.

Thermal behavior of fly ash based geopolymers

A visual assessment of the thermally treated samples was performed. All geopolymer specimens either pastes or mortars showed a significant change of color from black to dark grey and finally light brown. As temperature increased to more than 400°C, macro cracks were observed on the specimens' surface. However, the initial cracking temperature seems to be related to the presence of the sand as follows: over 600°C for geopolymer pastes and 400°C for sand to fly ash ratios equal to 1.0. In the case of geopolymer mortars, the initial cracking temperature is decreased as a result of thermal incompatibility between the geopolymer matrix and the sand.

Figure 3 presents the % mass loss of geopolymer pastes and mortars after their thermal treatment. As it can be seen, all geopolymer samples either pastes or mortars experienced considerable mass loss as temperature increases. Geopolymer pastes showed a sharp mass reduction of 25% at 300°C. Above 300°C and up to 600°C, mass loss continued to increase in a slower rate, while above this temperature the mass almost stabilized. The geopolymer mortars exhibit a similar behaviour up to 600°C, with lower values of mass loss, due to the lower content of geopolymeric matrix. Unlike geopolymer pastes, the geopolymer mortar specimens exhibit a rapid mass decline above 600°C, because of the starting dissociation of limestone. It must be noted that at 800°C the mass loss is the same for both pastes and mortars as the lower mass loss of geopolymeric matrix is counter balanced with the mass loss of aggregates.

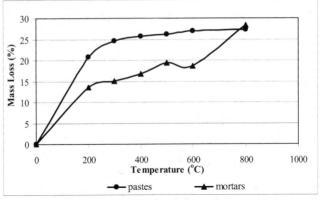

Figure 3. % Mass loss of specimens in relation to the exposure temperature.

Figure 4 presents the average volume reduction of geopolymer specimens in relation to the temperature of thermal treatment. Unlike the mass loss that is almost stabilized above 400°C, the shrinkage of the geopolymer specimens continues to increase, reaching an almost 20% of volume reduction at 800°C. There is a similar trend in the geopolymer mortar specimens but the shrinkage level is significally lower (approximately 7% volume reduction at 800 °C).

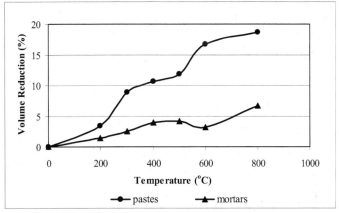

Figure 4. Volume reduction of specimens in relation to the exposure temperature

Figure 5 presents the compressive strength of the geopolymer pastes and geopolymer mortars after their thermal treatment. As it can be seen, the incorporation of aggregates lowers the initial compressive strength of geopolymers. The exposure of the mortars to elevated temperatures results to a continuous decrease of their compressive strength. The pastes retain 94% and 65% of their initial strength after thermal treatment at 200 and 500°C, respectively, but at higher temperature a rapid decline of the pastes' compressive strength is observed, leading to the total collapse of their structure at 800°C. It must be noted that mortar specimens at this temperature retain their cohesiveness.

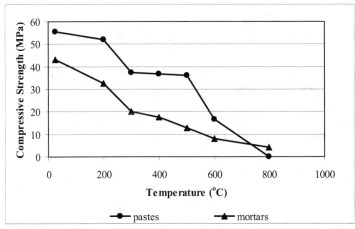

Figure 5. Compressive strength of specimens in relation to the exposure temperature

The ultrasonic pulse velocity was measured before and after each thermal treatment. The decrease of the ultrasonic pulse velocity indicates the formation of microcracks in the interior of the specimens. Figure 6 shows the ultrasonic pulse velocity of the specimens, in relation to the temperature of thermal treatment. The observed changes are in accordance with the mass loss and the strength measurements. The pastes showed a sharp decrease of ultrasonic pulse velocity up to 300°C, while the mortars exhibit a continuous decrease through the whole temperature range.

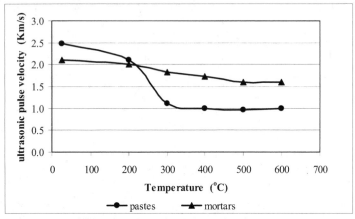

Figure 6. Ultrasonic pulse velocity in relation to the exposure temperature

Figures 7 and 8 present the XRD patterns of geopolymer pastes and geopolymer mortars, respectively, before and after their thermal exposure to 200°C, 400°C, 600°C and 800°C.

As it can be seen from Figure 7, the thermal treatment of pastes up to 600°C does not cause any significant changes as far as the mineral composition is concerned. The only observed difference is the gradual reduction of the hump between 20 and 35° which is attributed to the amorphous aluminosilicate product of the geopolymerization. The specimen treated at 800°C contains more gehlenite and the new aluminosilicate phase of nepheline. It seems that as the geopolymeric matrix decomposes, the aluminosilicate species react with the released alkalies and calcium to form crystalline phases (nepheline and gehlenite, respectively).

The XRD patterns (Figure 8) of the geopolymer/aggregate composites confirm the formation of gehlenite and nepheline after exposure at 800°C. This sample contains also portlandite coming from the decomposition of the calcareous aggregates.

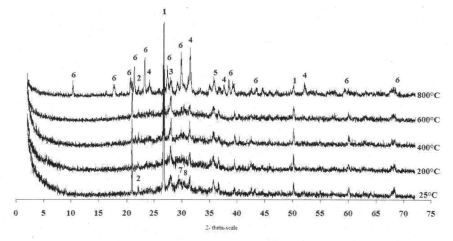

Figure 7: XRD patterns of geopolymer pastes in relation to the exposure temperature
(1: quartz, 2: cristobalite, 3: feldspars, 4: gehlenite, 5: maghemite, 6: nepheline, 7: natrite, 8: calcite)

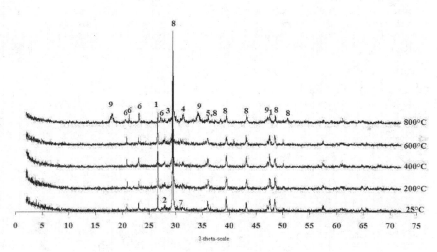

Figure 8: XRD patterns of geopolymer mortars in relation to the exposure temperature
(1: quartz, 2: cristobalite, 3: feldspars, 4: gehlenite, 5: maghemite, 6: nepheline, 7: natrite, 8: calcite,
9: portlandite)

All the above measurements are in good accordance with each other. The geopolymer pastes showed a high mass loss up to 300°C, accompanied by sharp shrinkage, internal microcracking and strength loss. All these indicate rapid migration and removal of bound water. After that, a more or less stabilized condition is observed up to 500°C, but at higher temperature the strength is sharply decreased and the specimens are totally collapsed at 800°C. XRD measurement confirmed that at high temperatures the amorphous geopolymeric matrix is gradually replaced with crystalline aluminosilicate phases and the consequent volume changes are probably the reason for the total collapse of the specimens.

The geopolymer binder in the geopolymer mortars obviously behaves in the same way as the geopolymer paste. The presence of the aggregates leads to the reduction of the shrinkage effect as well as the mass loss, leading to a smoother and continuous strength loss over the whole temperature range. However, the thermal incompatibility between the geopolymer matrix and the sand accelerates the surface cracking of the specimens.

CONCLUSIONS

The optimal conditions for the synthesis of geopolymers having as raw material fly ash originating from Megalopolis are: R/Al=1.0, Na/(Na+K)=1.0 and [Si]/R_2O=1.0 (R: Na,K). The parameter having the higher impact on the development of compressive strength is the kind of alkali ion. Higher strength is achieved when Na is used, while the gradual substitution of Na by K lowers the compressive strength. The second important parameter is the silicon content in the alkaline solution. The optimal value of this parameter is 1.0 and any further increase up to 2.0 induces a negative effect on the geopolymers' compressive strength. The R/Al ratio seems to have a marginal effect on the mechanical properties of the specimens.

Fly ash based geopolymers possess rather low thermal stability. The pastes retain 94% and 65% of their initial strength after thermal treatment at 200 and 500°C, respectively, but at higher temperature a rapid decline of the pastes' compressive strength is observed, leading to the total collapse of their structure at 800°C, where the initially amorphous structure was replaced by nepheline and gehlenite. The exposure of the geopolymer mortars to elevated temperatures results to a continuous decrease of their compressive strength.

REFERENCES
1 J. Davidovits, Geopolymers: Inorganic polymeric new materials, *J. Mater Edu*, **16**, 91-114 (1994)
2 J. Davidovits, Geopolymers: Inorganic polymeric new materials, *J. Therm Anal Calorim*, **37**, 1633-1656 (1991)
3 Hua Xu, J.S.J. van Deventer, The geopolymerisation of aluminosilicate materials, *Int J Miner Process*, **59**, 247-258 (2000)
4 Hua Xu, J.S.J. van Deventer, The effect of alkali metals on the formation of polymeric gels from alkali feldspars, *Colloids Surf A Physicochem Eng Asp*, **216**, 27-44 (2003)
5 Hua Xu, J.S.J. van Deventer, G.C Lukey, Effect of alkali metals on the preferential geopolymerisation of Stilbite/ Kaolinite mixtures, *Ind Eng Chem Res*, **40**, 3749-3756 (2001)
6 M. Rowles, B. O' Connor, Chemical optimisation of the compressive strength of aluminosilicate geopolymers synthesized by sodium silicate activation of metakaolinite, *J Mater Chem* **13(5)**, 1161-1165 (2003)
7 M. Schmucker, K.J.D. Mackenzie, Microstructure of sodium polysialate siloxo geopolymer, *Ceram Int*, **31**, 433-437 (2005)
8 J.W. Phair, J.S.J. van Deventer, Characterisaton of fly ash based geopolymeric binders activated with sodium aluminate, *Ind Eng Chem Res*, **41**, 4242-4251 (2002)

9   A.M. Fernandez- Jimenez, A. Palomo, M. Criado, Microstructure development of alkali activated fly ash cement: a descriptive model, *Cem Concr Res*, **35**, 1204-1209 (2005)

10 T. Backarev, Geopolymeric material prepared by using Class F fly ash and elevated temperature curing, *Cem Concr Res*, **35**, 1224-1232 (2005)

11 W.K.W. Lee, J.S.J. van Deventer, G.C LuKey, Effect of anions on the formation of aluminosilicate gel in geopolymers, *Ind Eng Chem Res*, **41**, 4550-4558 (2002)

12 Hua Xu, J.S.J. van Deventer, Geopolymerisation of multiple minerals, *Miner Eng*, **15**, 1131-1139 (2002).

13 J.C. Swanepoel, C.A. Strydom, Utilisation of fly ash in geopolymeric materials, *Appl Geochem* **17**, 1143-1148 (2002)

14 P.S. Singh, M. Trigg, I. Burgar, T. Bastow, Geopolymer formation processes at room temperature studied by Si and Al MAS- NMR, *Mate Sci Eng A*, **396(1-2)**, 392-402 (2005)

15 R.A. Flecher, C.L. Nicholson, S. Shimada, K.J.D. Mackenzie, The composition of aluminosilicate geopolymers, *J Eur Ceram Soc*, **25**, 1471-1477 (2005)

16 D. Feng, H. Tan, J.S.J. van Deventer, Ultrasound enhanced geopolymerisation, *J Mater Sci*, **39**, 571-580 (2004)

17 Duxson P., J L. Provis, G. C. Lukey and J. S.J. van Deventer (2007). The role of inorganic polymer technology in the development of 'green concrete, *Cem Concr Res*, **37(12)**, 1590-1597 (2007)

18 Hua Xu, J.S.J. van Deventer, Effect of source materials on Geopolymerisation, *Ind Eng Chem Res*, **42**, 1698-1706 (2003)

19 Panagiotopoulou Ch., Perraki T., Tsivilis S., Skordaki N., Kakali G., A study on alkaline dissolution and geopolymerisation of hellenic fly ash, *Ceramic Engineering and Science Proceedings*, **29(10)**, 165-174, (2009).

20 Komnitsas K, Zaharaki D., Geopolymerisation: A review and prospects for minerals industry, *Min Eng*, **20 (14)**,1261-1277 (2007)

21 Komnitsas K, Zaharaki D, Perdikatsis V, Effect of synthesis parameters on the compressive strength of low- calcium ferronickel slag inorganic polymers, *J Haz Mat*, **161 (2-3)**, 760-768 (2009)

22 J. Davidovits, Structural Characterisation of Geopolymeric Materials with X-Ray Diffractometry and MAS-NMR Spectrometry, *Proccedings, Geopolymer* **88**, 149-166 (1998)

23 V.F.F. Barbosa, K.J.D. Mackenzie, C.D. Thaumaturgo, Synthesis and Characterisation of sodium polysialate inorganic polymer based on alumina and silica, *Proccedings, Geopolymer* **99**, 65-78 (1999)

24 Lee W.K.W, J.S.J. Van Deventer J.S.J. The effects of inorganic salts contamination on the strength and durability of geopolymers, Colloids and Surfaces A: Physicochemical Engineering Aspects **211(1)**, 115-126 (2002)

25 P. Duxson, J.L. Provis, G.C. Lukey, S.W. Mallicoat, W.M. Kriven, J.S.J. van Deventer, Understanding the relationship between geopolymer composition, microstructure and mechanical properties, *Colloids Surf A Physicochem Eng Asp*, **269**, 47-58 (2005)

26 Duxson P., Lukey G.C., Separovic F, Van Deventer J.S.J, Modelling speciation in highly concentrated alkaline silicate solutions, *Ind Eng Chem Res*,**44** (4), 8899-8908 (2005)

27 Bakharev T. Thermal behaviour of geopolymers prepared using class F fly ash and elevated temperatures curing. *Cem Concr Res,* **36**, 1134 -1147 (2006)

28 Kong D.L.Y., Sanjayan J.G. Effect of elevated temperatures on geopolymer paste, mortar and concrete, *Cem Concr Res,* **40**, 334-339 (2010)

29 Davidovits J. Geopolymeric reactions in the economic future of cements and concretes: world-wide mitigation of carbon dioxide emission, In *Proceedings* of the 2nd International Conference on Geopolymer '99, Saint Quentin, France, 30 June-2 July, 111- 121 (1999)

30 Li Z., Ding Z, Zhang Y. Development of sustainable cementitious materials, *Proceedings of International Workshop on Sustainable Development and Concrete Technology*, Beijing, China, 55-76 (2004)

31 Ch. Panagiotopoulou, Synthesis and properties of geopolymers deriving from industrial minerals and by-products, *pHD thesis, NTUA,* Athens, (2009)

32 Panagiotopoulou Ch., Kontori E., Perraki Th., Kakali G., Dissolution of aluminosilicate minerals and by-products in alkaline media. *J Mat Sci*, **42**, 2967-2973 (2007)

33 Hua Xu, Geopolymerisation of aluminosilicate minerals, *pHD Thesis, The University of Melbourne* (2002)

34 C.K.Yip, G.C.Lukey, J.S.J. van Deventer, The coexistence of geopolymeric gel and calcined silica hydrate at the early stage of alkaline activation, *Cem Concr Res*, **35(9)**, 1688-1697 (2005)

MECHANICAL RESPONSE OF DISCONTINUOUS FILAMENT PVA FIBER REINFORCED
GEOPOLYMERS

Benjamin Varela[1] and Jeffrey W. Rogers[2]

[1]Mechanical Engineering Department
[2]Civil Engineering Technology Department
Rochester Institute of Technology
Rochester, NY, USA

ABSTRACT
    Geopolymers are an emerging class of ceramic materials that have great potential in several industries. The scientific literature reported significant benefits when using fibers such as glass, carbon, and boron as reinforcement for geopolymer mixtures. The use of these fibers within a geopolymer mixture will add significant benefits through increased ductility, flexibility, and impact strength. However, in high alkalinity environments, these benefits are significantly reduced.
    The paper will prove that Poly-Vinyl Alcohol (PVA) microfibers are stable under high alkaline conditions, which will improve the compressive strength, bending strength and ductility of Geopolymers. The goal of this paper is to improve the mechanical response of geopolymers by using low cost, high strength, low density PVA microfibers. The objectives of this paper are to improve: (1) compressive strength; (2) bending strength; and (3) ductility of the geopolymer paste by adding 0.5%, 1.0% and 1.5% per total volume of PVA microfibers. This paper describes how the geopolymer paste was synthesized using fly ash, how the fibers were incorporated, the mechanical testing performed and the results obtained.
    This work demonstrates that RECS7 microfibers yielded the best compressive and bending strengths. This work also reveals that the best ductility was obtained using the RECS15 microfibers.

INTRODUCTION
    Geopolymers are an emerging class of ceramic materials that have great potential in several industries[1,2]. During the past years several research groups have explored their properties for a possible alternative to Ordinary Portland Cement and for the development of a new green cementating material. It is known that geopolymers can be synthesized from heat-treated kaolin clay or from byproducts such as fly ash and blast furnace slag.
    Geopolymer pastes show some desirable engineering properties such as high early compressive strength, thermal stability, low curing temperature and resistivity to acid attack[3]. However, as is the case with many ceramic materials, geopolymer pastes are brittle. To over come this deficiency the field of Geopolymers has followed many of the techniques used in the concrete industry. For example, cement is mixed with different aggregates and fibers to overcome brittleness. In this paper microfibers were mixed into geopolymer paste to enhance its ductility. In the case of fibers, is generally accepted that the high alkalinity of the paste can damage the fibers, thus hindering the types of fibers that can be incorporated.
    The hypothesis of this paper will prove that Poly-Vinyl Alcohol (PVA) microfibers are stable under high alkaline conditions, which will improve the compressive strength, bending strength and ductility of Geopolymers. This preliminary research looks at improving the mechanical response of geopolymers by using low cost, high strength, low density PVA microfibers. The objectives of this paper are to improve: (1) compressive strength; (2) bending strength; and (3) ductility of the geopolymer paste by adding 0.5%, 1.0% and 1.5% per total volume of PVA microfibers.
    This paper describes how the geopolymer paste was synthesized using fly ash, how the fibers were incorporated, the mechanical testing performed and the results obtained.

BACKGROUND

According to the American Coal Ash Association[4] 136 million tons of Coal Combustion Residues (CCR) were produced in the US in 2008. Approximately 50% of this amount corresponds to Fly Ash. CCR are a significant environmental problem and researchers are looking for beneficial uses for these residual materials. For example blending fly ash with clinker has proven successful as an alternative raw material in the Portland Cement Industry. This technique allows up to 70% substitution of clinker content without adverse effects in the performance of the cement[5]. Today geopolymer researchers are exploring fly ash as the primary raw material for the synthesis of Geopolymer pastes and concretes, especially for large scale applications. In this research F type Fly Ash was donated by Boral Materials Technology and used as the primary source of reactive alumina and silica for the geopolymer paste. F type fly ash was selected for this research because it proved to be highly reactive and low cost with manageable rheology.

This research uses PVA microfibers as reinforcement for the geopolymer paste because they have low density, high tensile strength and high modulus of elasticity as shown in Table 1.

Table 1. Mechanical properties of PVA fibers

Fiber	Diameter (mm)	Cut length (mm)	Specific Gravity	Tensile Strength, GPa (Ksi)	Modulus of Elasticity, GPa (Ksi)
RECS7	0.027	6	1.3	1.6 (232)	39 (5656.4)
RECS15	0.04	8	1.3	1.6 (232)	40(5801.5)

Source: http://www.kuraray-am.com/pvaf/fibers.php

Additionally, PVA fibers tend to develop very strong chemical bonds due to the presence of hydroxyl groups in its molecular chains[6]. This characteristic to form strong bonds without any further treatment of the microfibers may be desirable under the high alkali conditions normally found in geopolymer pastes. The use of PVA microfibers in concrete materials has been pioneered by Victor C. Li. He found that by adding up to 2% by total volume of these microfibers ductility was increased by approximately four fold in reinforced concrete load bearing members[7]. This improvement in ductility is created by the crack bridging effect of the randomly dispersed fibers.

METHODOLOGY

In this work F type fly ash, Na base alkaline solution and two types of PVA microfibers were used as raw materials. The samples were prepared by adding the microfibers into the reactant mix such that their volumes represented 0.5%, 1.0% and 1.5% of the total volume. The mechanical responses of the samples were tested after 28 days. This procedure is described in detail in the following sections.

Raw Materials

Class F type fly ash from Boral Material Technologies was used as the primary source of reactive silica and alumina. Table 2 presents the chemical composition of the fly ash used as provided by the supplier.

Table 2. Chemical composition in terms of oxides of Boral Material Technologies high performance Fly Ash as provided by supplier.

Oxide	SiO_2	Al_2O_3	Fe_2O_3	K_2O	CaO
Percentage	55.4	27.6	7.2	2.7	1.2

An alkali activating solution was prepared by using E type sodium silicate (PQ Corporation), a 40% NaOH solution (Brainerd Chemical Company) and distilled water. The ingredients were mixed and slowly stirred to yield a solution with molar ratios of SiO_2/Na_2O of 1.2 and H_2O/Na_2O of 14, after this the solution was left to mature for 1 hour. Two types of Poly-Vinyl Alcohol (PVA) fibers (Kuraray America) named RECS7 and RECS15 as described in Table 1 were used to enhance the mechanical response of the geopolymer. Figure 1 presents the physical differences and dimensions between both types of fibers.

Figure 1. Top picture depicts the RECS7 fiber and Bottom picture depicts the RECS15 fiber.

Sample Preparation

To prepare the base samples F Type fly ash was slowly added to the activating solution until a 2:1 mass ratio was achieved using a high shear mixer until the paste looks homogenous. The paste was then poured into the 1x1x6 inch molds and vibrated for 10 minutes.

To prepare the reinforced samples the F type fly ash was mixed with the PVA microfibers in a rotary mill for 10 minutes and then followed the procedure mentioned above. The volume of the PVA microfibers added represented 0.5 %, 1.0% and 1.5% of the total volume of the reinforced geopolymer.

The samples were left to mature at room temperature for 1 hour. After this, the samples were cured using steam for 24 hours. After curing, the samples were removed from the molds and let to mature in the laboratory for 28 days, after which they were mechanically tested.

Mechanical testing

A MTS model 451G Universal Testing machine was used for mechanical testing. For compressive testing the 1x1x6 inch molded beams were cut into cubes of 1x1x2 inch. For the bending testing, a 1x1x6 inch molded beam was placed in a three point bending fixture leaving a 4 inch span between the adjustable supports. The speed of the crosshead was kept constant at 0.02 in/min for all tests. Since this is a preliminary study this work used small size specimens, therefore ASTM standards were not followed.

RESULTS AND DISCUSSION

Figure 2 presents the results of the compressive and bending tests for the base geopolymer samples. Three beams were used for bending testing while three cubes were used for compressive testing.

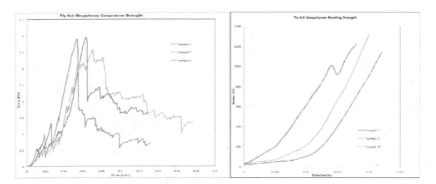

Figure 2. The left figure depicts the compressive stress-strain diagrams while the right figure depicts the bending stress-deflection diagramfor the three samples of base geopolymer.

Figure 2 shows that the average compressive strength is 3900 Psi (26.9 MPa) and the average bending strength is 1220 Psi (8.4 MPa). The failure mode of the bending samples was observed to be similar to that of brittle materials.

Figure 3 presents the average results of the adition of the PVA fibers to the base samples. Three samples were used to obtain these averages.

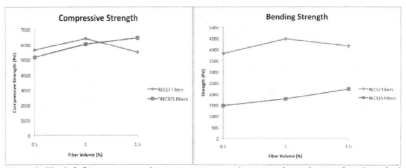

Figure 3. The left figure presents the average compressive strength results as a function of microfiber volumes. The right figure presents the bending strength as a function of microfiber volumes.

Figure 3 reveals that the average compressive strength in the samples with RECS7 microfibers decreased from 6,380 Psi (43.9 MPa) to 5,480 Psi (38.8 MPa) when the microfiber volume was increased from 1.0 % to 1.5% of the total volume. The samples with RECS15 microfibers did not display this annomaly, which continued to increase to 6,434 Psi (44.3 MPa) at 1.5% of the total volume. The researchers believe that this effect might have been caused by the difficulty associated with homogeneusly mixing the RECS7 microfibers at 1.5% volume with the available equipment. Additionally the researchers believe that differences in surface areas at this volume played a significant role in the mechanical response. This anomaly will be investigated in future studies.

Figure 3 also reveals that the average bending strength in the samples with RECS7 microfibers was significantly higher than the samples with RECS15 microfibers reaching a maximum of 4,470 Psi (30.8 MPa) at 1.0% volume. As previously stated, the researchers were unable to verify if this mechanical response was caused by the inhomogeneity of the microfibers in the samples. This will be investigated in future studies.

Figure 4. Fracture surface of a microfiber reinforced Geopolymers at low magnification (left) and high magnification (right).

Figure 4 reveals that at low magnification there exists a slanted micro-crack, which is associated with ductile materials. The microfibers are shown bridging the micro-crack. At high magnification one can see that the microfibers are not in tension and did not break during the passing of the micro-crack.

Figure 5 shows that a ductile material was developed by the addition of the microfibers.

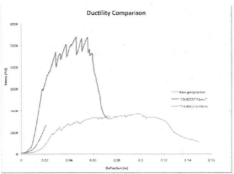

Figure 5. Ductility comparison among base Geopolymers, Geopolymers with 1% of RECS7 fibers and Geopolymers with 1% of RECS15 fibers using a three point bending test.

Figure 5 reveals that a ductile material was achieved in the sample using the RECS15 microfibers. This represents an enhancement in ductility of approximately 5 times compared with the base Geopolymers sample. The stress-strain response for the sample using the RECS7 microfibers shows a similar saw-tooth behavior found by Li in PVA microfiber reinforced concrete[8].

CONCLUSIONS

This preliminary research evaluated the mechanical response of geopolymers by using low cost, high strength, low density Poly-Vinyl Alcohol (PVA) microfibers. This work proves that PVA microfibers are stable under high alkaline conditions, which improved the compressive strength, bending strength and ductility of Geopolymers.

This work shows that the best compressive and bending strengths were obtained with the RECS7 microfibers. This work also reveals that the best ductility was obtained using the RECS15 microfibers.

Future work will include a study of the dispersion of the microfibers within the Geopolymers and the evaluation of the bonding characteristics of the microfibers to the Geopolymers.

ACKNOWLEDGEMENTS

The authors are grateful for the assistance of the following corporations who contributed in this research: Boral Material Technologies from Atlanta GA and Kuraray America. The authors are also grateful for the assistance provided by the USAF Office of Scientific Research.

REFERENCES
1.-J. Davidovits, Geopolymer chemistry and applications".Saint-Quentin, Institut Géopolymère, c2008.
2.-J. Davidovits, Geopolymers, inorganic polymeric materials, Journal of Thermal Analyis,**37**, 1633-1656 (1991).
3.-P. Duxson, A. Fernández-Jiménez, J. Provis, G. Lukey, A. Palomo, and J. Deventer 2 Geopolymer technology: the current state of the art, Journal of Materials Science **42**, no. 9: 2917-2933 (2007).
4.http://acaa.affiniscape.com/associations/8003/files/2008_ACAA_CCP_Survey_Report_FINAL_100 509.pdf, accesses on December 31st, 2010.
5.- M. Phelan, 2008. Chemistry for construction, Chemical Engineering **115**, no. 6: 22-5 (2008).
6.- V. C. Li, C. Wu, W. Shuxin, O. Atsuhisa, S. Tadashi, Interface tailoring for strain-hardening Polyvinyl Alcohol- engineering cementitious Composites (PVA-ECC), ACI Materials Journal, 463-472 (2002).
7.-V. C. Li,Large volume, high performance applications of fibers in civil engineering, Journal of Applied Polymer Science,**83**, 660-686 (2002).
8.- J. H. Yu, Research on production, performance and fibre dispersion of PVA engineering cementitious composites, Materials Science and Technology, **25**, 651-656 (2009).

MICROWAVE ENHANCED DRYING AND FIRING OF GEOPOLYMERS

Tyler A Gubb, Inessa Baranova, Shawn M. Allan, Morgana L. Fall, Holly S. Shulman
Ceralink Inc.
Troy, New York, USA

Waltraud M. Kriven
University of Illinois at Urbana-Champaign
Urbana, Illinois, USA

ABSTRACT

The feasibility of using microwave energy to dry and fire pre-cured geopolymers was experimentally demonstrated, and supported by analysis of published microwave dielectric data for geopolymers. Dielectric loss tangents and half power microwave absorption depths were calculated from published room temperature dielectric constant and loss values of various geopolymer compositions. The published data indicated that geopolymers would heat at room temperature with microwave energy. Several laboratory experiments were performed to test the heating behavior of sodium and potassium based geopolymer compositions. Experiments demonstrated more vigorous microwave heating with sodium geopolymers than with potassium geopolymers. Both compositions were dried in less than 10 minutes with pure microwave heating. Further heating with pure microwave energy resulted in non-uniform, rapid heating, or "thermal runaway", with localized melting of the geopolymer. Hybrid microwave heating with susceptors resulted in uniformly fired geopolymers, without melting.

INTRODUCTION

Interest in geopolymers has accelerated in recent years due to their possible use in structural applications, dentistry, and hazardous waste stabilization, while maintaining a trivial environmental impact compared to traditional building materials. Many useful geopolymer compositions are fabricated by low temperature chemical reaction based curing processes[1-3]. Some recent research has been reported for using microwave energy to enhance curing and drying[4, 5]. Other research has focused on high temperature firing of geopolymers, which can develop stronger glass-ceramic materials[6-8]. In situations where heat is required, as in drying and firing, microwave energy provides an energy efficient alternative in place of conventional heating. Traditional radiant heating methods rely on thermal conduction to deliver heat throughout a material. Microwaves generate heat throughout the volume of the material, allowing faster, more uniform heating to occur. Volumetric microwave heating helps to overcome sluggish endothermic phase transitions, such as evaporation of water or decomposition of kaolin to metakaolin through the loss of hydroxides. Both drying and dehydration occur in the firing of geopolymers. In traditional heating, these reactions often require slow heating, as the endothermic reaction prevents heat from progressing into the product until the reaction completes first at the surface. Microwaves can generate heat throughout the part despite an endotherm[9].

When microwave energy is the sole source of heat, a material that absorbs microwave energy will heat volumetrically, but cool from the surface. This situation creates an "inverse temperature profile" in which the sample is warmer inside and cooler at the surface during heating. This is opposite of traditional radiant heating where the material will be cooler in the center. In some materials, the inverse temperature profile can lead to thermal runaway – where the hotter center heats better than the cooler surface, and in turn heats better in the microwave. The thermal runaway can lead to molten centers with unfired surfaces.

The most practical way to prevent thermal runaway is through hybrid microwave heating. Hybrid heating combines a radiant heat source with microwave energy, providing a uniform temperature profile throughout the sample as heat is generated at the interior while heat conducts from the exterior. This combination results in a uniform temperature profile and in turn improved properties. Two types of hybrid heating are susceptors, and Microwave Assist Technology (MAT). Susceptors function like wireless microwave heating elements, efficiently converting microwave energy into heat. When used with insulation thermal packages, susceptors can be used to fire ceramics even in standard kitchen microwaves[10]. MAT is based on traditional kilns and uses gas or electric radiant heat, which is controlled independently of the microwave energy[11]. MAT generally requires less microwave power, which reduces equipment costs and simplifies scale-up[12].

Dielectric Properties

The effectiveness of microwave heating for materials is generally determined by the dielectric properties at the microwave frequency. Two values represent the dielectric properties of a material, dielectric constant, ε', and dielectric loss, ε''. Dielectric constant represents the ability for ions and dipoles within a material to polarize in response to an alternating electric field, and also determines the wavelength of the microwave energy within the material. The dielectric loss represents the degree to which an alternating field is converted to heat energy. From these two values can be derived the loss tangent, tan δ, and the half-power depth. A loss tangent between 0.01 and 1 generally indicates that a material will heat well with microwave energy. Below 0.01, materials tend to be microwave transparent, while above 1 materials become reflective[13]. The half power depth measures the distance at which 50% of the microwave energy passing through a material is dissipated. The equations for loss tangent (1) and half power depth (2) are expressed below. In equation 2, D_{HP} is the half power depth, c is the speed of light, ω is the angular frequency ($2\pi f$), and ε_o is the permittivity of free space.

$$Tan(\delta) = \frac{\varepsilon''}{\varepsilon'} \tag{1}$$

$$D_{HP} = \frac{c(\ln 2)}{2\omega\varepsilon_0}\left(\frac{2}{\left(\sqrt{1+(\tan\delta)^2}-1\right)\varepsilon'}\right)^{1/2} \tag{2}$$

Published room temperature dielectric data was obtained from Jumrat et al[5]. From this data, loss tangent and half-power depth could be calculated to determine support for microwave heating from room temperature. Geopolymers are composed significantly of both free water and hydroxides surrounding charge balancing alkali cations. These structures suggest strong likelihood for geopolymers to heat well with microwave energy, with mobile water dipoles and ions. The Jumrat paper presented just ε' and ε'', and data was measured as a function of curing time. As the geopolymer cured, the dielectric loss and dielectric constant both decreased, suggesting reduced mobility of polar groups as curing progressed. Ceralink selected two of the data sets to calculate the loss tangent and half power depths. These calculations strongly supported microwave heating of geopolymers from room temperature (Figure 1).

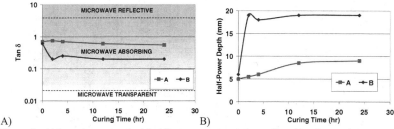

Figure 1: A) Loss tangents calculated for two representative sodium-based geopolymer compositions reported by Jumrat, et al. B) Half-power depth for the two geopolymer compositions.

The loss tangent values determined by the present authors from Jumrat's data were between 0.1 and 1.0, indicating that these geopolymer materials will heat very well in microwave, at least from room temperature. The half-power depth indicates that the microwave energy transmits farther into the material as curing completes. This heating is likely due to the presence of water molecules within the crystal lattice and the charge balancing cations, in this case sodium. At elevated temperatures after drying, the cation is likely the greatest contributor to suspecting action, as water and hydroxides are thermally removed from the geopolymer structure by approximately 400 °C, as shown by Bell, et al[7]. In order to understand microwave heating above room temperature, high temperature dielectric measurements will be required. These will be investigated in future studies.

This study investigated experimental microwave heating of geopolymers with sodium and potassium charge balancing cations. The effect of the alkali cation used in the geopolymer was studied through the laboratory experiments. Microwave firing of a potassium based geopolymer with basalt fiber reinforcement was also studied. Microwave heating was explored with microwave as the only source of heat to the geopolymer, and also with a hybrid susceptor-based method.

EXPERIMENTAL PROCEDURE

Pre-cured sodium and potassium-based geopolymer samples were weighed and dimensions measured before being placed in an alumina fiber thermal package for heating experiments. Some of the potassium-based geopolymer samples were reinforced with 7 wt % chopped basalt fibers. The thermal package was made of 2.5 cm thick Unifrax Duraboard 2600® insulation cylinder, with 4 cm thick base and lid, and 8 cm inner diameter. The geopolymer sample was placed on a bed of alumina powder in an alumina crucible in the center of the thermal package. The thermal package allowed room for adding two 25 gram silicon carbide (SiC) susceptors (Research Microwave Systems "Thermcepts") for hybrid susceptor-assisted heating studies. Figure 2 (A and B) shows schematics of these set-ups without and with susceptors, respectively, compared to photographs of the set-ups as used for experimentation (Figure 2 C and D). The geopolymer composites here examined were prepared and studied by Rill et al. and the findings were published elsewhere.[14]

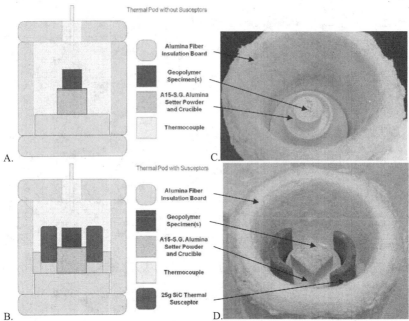

Figure 2. Schematics of A) Alumina fiber thermal package without suscepting material, used for "pure" microwave heating. B) Alumina fiber hybrid thermal package with two 25 g SiC susceptors positioned on either side of sample and setter. C) Photograph of thermal package without SiC susceptors. D) Photograph of thermal package with susceptors. Both set-ups used A15-S.G. alumina powder contained within an alumina crucible as a setter. A thermocouple was extended through the lid, and positioned approximately 1.5 cm above the sample.

Four experiments (Table I, 1-4) were performed to study the feasibility of drying sodium and potassium based geopolymers via direct heating, i.e. without the use of suscepting material (Figure 2A and C). Samples were arranged within a Research Microwave Systems ThermWave 1.3 microwave system (2.45 GHz, 900 W maximum power) chamber and microwave energy was applied for 15 minutes, ranging from 20% power (180 W) to 50% (450 W), with the same power-time profile for each sample. Temperature within the chamber was recorded via the thermocouple. Once the samples were cooled, final weight and dimensions were measured.

Four experiments were used to study the firing of potassium-based geopolymer with basalt fiber reinforcement. Direct microwave heating was used in one experiment, while experiments 6-8 employed SiC susceptors within the thermal package (Figure 2B). Target firing temperatures for samples 6, 7, and 8 were 800, 900, and 1000 °C, respectively and were weighed, measured, and cross-sectioned vertically to observe the interior structure once cooled. The temperature within the chamber was recorded via the thermocouple. An optical pyrometer with an emissivity setting of 0.90 was used after heating to determine the actual sample temperature after the microwave power was stopped. Sample descriptions are provided below in Table I.

Table I: Descriptions of geopolymer (GP) microwave heating experiments for drying and firing sodium and potassium based geopolymers.

Experiment No.	Experiment Purpose	Geopolymer Sample Composition	Geopolymer Mass (g)	Heat Treatment Temperature (°C)
1	Drying, self-heating	Na-based geopolymer	2	114
2	Drying, self-heating	Na-based geopolymer	16	209
3	Drying, self-heating	K-based geopolymer	16	143
4	Drying, self-heating	K-based geopolymer, Basalt fiber reinforced	16	123
5	Firing, self-heating	K-based geopolymer, Basalt fiber reinforced	16	560
6	Firing with susceptors	K-based geopolymer Basalt fiber reinforced	16	800
7	Firing with susceptors	K-based geopolymer Basalt fiber reinforced	16	900
8	Firing with susceptors	K-based geopolymer Basalt fiber reinforced	16	973

RESULTS AND DISCUSSION

Geopolymer Microwave Drying Experiments

Heating curves for drying experiments are provided in Figure 3. The drying experiments revealed the difference in heating ability for sodium and potassium based geopolymers. The sodium composition heated much faster than the potassium based sample of equal mass, reaching over 70 °C higher temperature in the 15 minute experiment. This may be due to a higher mobility of the smaller sodium ion than the potassium ion. Dielectric property measurements of these two compositions would be extremely interesting to determine how these cations affect the properties of the geopolymers.

Another effect observed in the drying experiments was a mass effect, in which 16 grams of sodium geopolymer, heated the thermal package much more than a 2 gram sample. This mass effect is unsurprising, as the geopolymer was the only source for heating in the thermal package volume.

The basalt fiber reinforced, potassium based geopolymer heated more slowly than the unreinforced sample. Basalts have a wide range of chemical compositions, and therefore a range of microwave heating ability. In this case, the basalt appeared to heat less than the geopolymer, resulting in the slower microwave heating.

The sodium geopolymers had an average mass loss of 12%, while the unreinforced potassium geopolymers lost 13%. The reinforced potassium geopolymer, meanwhile, lost only 5.4% mass, despite reaching higher temperatures than the unreinforced sodium sample. Minor cracking was observed in some of the dried samples, however all samples remained intact.

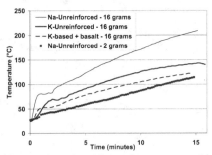

Figure 3: Heating curves for samples 1-4. All samples were ramped uniformly from 20% to 50% power. These temperatures were measured by the thermocouple positioned 1.5 cm above the geopolymer sample.

Geopolymer Microwave Firing Experiments
    The microwave heating profiles from the firing experiments of K-based geopolymers with 7 wt% basalt fiber reinforcement are provided in Figure 4. One sample was heated with microwave-only heating. This sample reached 200 °C in approximately 20 minutes, after which the heating rate accelerated quickly for 1-2 minutes prior to an arc. The electrical discharge (arc) provided a basis for terminating the experiment. The remaining samples were heated with susceptors to 800, 900, and 1000 °C, without any incidence of arcing.

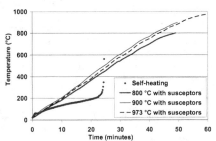

Figure 4: Heating profiles for the microwave firing experiments. All samples were K-based with 7% basalt-fiber reinforcement.

    The self heated sample only registered temperature at the thermocouple of up to approximately 550 °C immediately prior to arcing. However, upon examination of the sample (Figure 5), it was found that significant thermal runaway had occurred. The interior of the sample had melted, resulting in glassy bubbles and flow of the molten geopolymer, down into the alumina setter-powder. The molten region indicated that temperatures in excess of 1160 °C (the highest temperature achieved on susceptor fired materials, at which no melting was observed) had been achieved within the sample. Despite this intense internal heating, the top and side surfaces of the sample appeared as though they had not been fired. A cross-section of this sample revealed a dramatic example of an inverse temperature profile, leading to thermal runaway. This experiment lead to speculation that the dielectric loss and loss tangent of this geopolymer increases substantially as temperature increases. A shift to higher dielectric loss with increasing temperature would direct more microwave energy into the hotter material. The hotter regions, then would heat faster, meanwhile the less absorbing, cooler sample surfaces lose heat to the cold environment in the thermal package. This situation of "thermal runaway" has been observed and reported by several authors with microwave heating of many other materials systems[15-17]. The literature on thermal runaway supports the assumption of increasing dielectric loss, as well as the use of hybrid heating (e.g. susceptors) to prevent runaway. High temperature measurements of the dielectric properties of these materials are planned as future work to confirm this assumption. While melting and a non-uniform cross section were not desired, this experiment strongly demonstrated the feasibility of rapidly heating geopolymers to firing and melting temperatures using microwave heating.

It is also worthwhile to point out that geopolymers are nanoporous having 41 vol % porosity, with an average pore size of 3.4 nm radius and the porosity is closed rather than interconnected[18, 19].

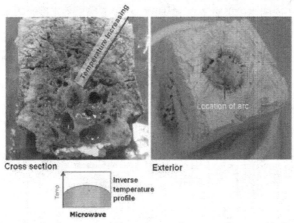

Figure 5: Photograph of microwave self-heated potassium geopolymer with 7 wt% basalt fiber reinforcement, after cross sectioning. The cross section shows dramatic evidence of the thermal runaway, in which the center of the sample melted and flowed down into the alumina powder bed. The exterior view of the sample shows essentially unfired geopolymer over most of the surface, with bulges at the centers of the sample faces resulting from the bubbles in the molten geopolymer. A schematic of the inverse temperature profile is shown below the cross section photograph.

The remaining samples 6-8 were fired with susceptors, in order to prevent thermal runaway. Hybrid firing with susceptors resulted in significantly more uniform cross sections than were seen with pure-microwave heating. Furthermore, the susceptor heated samples did not appear to contain any molten regions. Sample 8 was fired to the highest temperature, reaching 973 °C at the thermocouple. The sample temperature was measured to be 1160 °C by the optical pyrometer, immediately after the end of the microwave power application. This sample appeared to have the least cracking, the greatest shrinkage, and was noted to be more resistant to cutting than the samples fired at lower temperatures. The cross section and exterior of the susceptor fired geopolymer sample are shown below in Figure 6.

A)   B)

Figure 6: Photograph of A) exterior and B) cross section of Sample 8 after hybrid microwave-susceptor firing to a surface temperature of 1160 °C.

CONCLUSION
        Loss tangent and half-power depth where shown from the dielectric constant and loss of fly-ash based geopolymers. These dielectrics support the use of microwave energy for heating geopolymers to significant temperatures, as successfully demonstrated by laboratory experiments. Four geopolymer compositions were heated with pure microwave energy and achieved temperatures over 100 °C in less than 10 minutes time. In comparing geopolymers with similar mass but varying cation, an ion effect was observed where the smaller sodium ion resulted in faster heating than did potassium. Pure microwave energy was utilized to rapidly heat a K-based, basalt-reinforced geopolymer to significantly above its glass-ceramic transition temperature, with an inverse temperature profile. The dielectric properties of the K-based geopolymer were predicted to increase rapidly and dramatically above ~500 °C. Three additional samples of the same composition were fired to thermocouple temperatures of 800, 900, and 973 °C with the inclusion of susceptors to provide a radiant heat source. Hybrid heating was shown to produce uniform temperatures throughout the samples, preventing thermal runaway and resulting in uniformly appearing samples.

ACKNOWLEDGMENTS
The work of Waltraud M. Kriven was supported by the AFOSR under grant number 919 AF FA 9550-09-1-0322.

REFERENCES
[1]W. M. Kriven, J. L. Bell, and M. Gordon, Microstructure and Microchemistry of Fully-Reacted Geopolymers and Geopolymer Matrix Composites, *Ceramic Transactions*, (2003).
[2]A. Hajimohammadi, J. L. Provis, and J. S. J. Van Deventer, One-Part Geopolymer Mixes from Geothermal Silica and Sodium Aluminate, *Ind. Eng. Chem. Res.*, **47**, 9396-405, (2008).
[3]P. Duxson, J. L. Provis, G. C. Lukey, S. W. Mallicoat, W. M. Kriven, and J. S. J. Van Deventer, Understanding the Relationship between Geopolymer Composition, Microstructure and Mechanical Properties, *Colloids and Surfaces A: Physiochem. Eng. Aspects*, **269**, 47-58, (2005).
[4]J. Somaratna, D. Ravikumar, and N. Neithalath, Response of Alkali Activated Fly Ash Mortars to Microwave Curing, *Cement and Concrete Research*, **40**, 1688-96, (2010).
[5]S. Jumrat, B. Chatveera, and P. Rattanadecho, Dielectric Properties and Temperature Profile of Fly Ash-Based Geopolymer Mortar, *International Communications in Heat and Mass Transfer*, **38**, 242-48, (2011).
[6]P. He, D. Jia, T. Lin, M. Wang, and Z. Y., Effects of High-Temperature Heat Treatment on the Mechanical Properties of Unidirectional Carbon Fiber Reinforced Geopolymer Composites, *Ceramics International*, **36**, 1447-53, (2010).

[7]J. L. Bell, P. E. Driemeyer, and W. M. Kriven, Formation of Ceramics from Metakaolin-Based Geopolymers: Part 1-Cs-Based Geopolymer, *J. Am. Ceram. Soc.*, **92**, 1-8, (2009).

[8]N. Xie, J. L. Bell, and W. M. Kriven, Fabrication of Structural Leucite Glass-Ceramics from Potassium-Based Geopolymer Precursors, *J. Am. Ceram. Soc.*, **93**, 2644-49, (2010).

[9]S. Allan, M. Fall, H. Shulman, and G. Carnahan, Microwave Assist Sintering of Porcelain Insulators with Large Cross Section, *http://ceralink.com/publications/Daytona-Porcelain-011409.pdf*, (2009).

[10]G. Gaustad, J. Metcalfe, H. Shulman, and S. Allan, Susceptor Investigation for Microwave Heating Applications, *Innovative Processing and Synthesis of Ceramics, Glasses and Composites VIII*, **166**, (2005).

[11]J. Wang, J. Binner, and b. Vaidhyanathan, Evidence for the Microwave Effect During Hybrid Sintering, *Journal American Ceramic Society*, **89**, 1977-84, (2006).

[12]H. S. Shulman, and D. Spradling, Four for Foam, *American Ceramic Society Bulletin*, **87**, 22-24, (2008).

[13]H. S. Shulman, M. Fall, and S. Allan, Microwave Assist Technology for Product Improvement and Energy Efficiency, *Jeju, South Korea*, 142-46, (2007).

[14]E. Rill, D. R. Lowry, and W. M. Kriven, Properties of Basalt Fiber Reinforced Geopolymer Composites, *Cer. Eng. Sci. Proc.*, **31**, 57-69, (2010).

[15]P. S. Apte, and W. D. MacDonald, Microwave Sintering Kilogram Batches of Silicon Nitride, *Microwaves III*, 55-62, (1995).

[16]H. Fukushima, G. Watanabe, and M. Matsui, Microwave Sintering of Electronic Ceramics, *Journal of the Japan Society of Precision Engineering/Seimitsu Kogaku Kaishi*, **58**, 75-80, (1992).

[17]M. J. Ward, Thermal Runaway and Microwave Heating in Thin Cylindrical Domains, *IMA Journal of Applied Mathematics*, **67**, 177-200, (2002).

[18]J. L. Bell, and W. M. Kriven, Nanoporosity in Geopolymeric Cements, *Microscopy and Microanalysis '04: Proc. 62nd Annual Meeting of Microscopy Society of America*, **10**, 590-91, (2004).

[19]W. M. Kriven, M. Gordon, and J. L. Bell, Geopolymers: Nanoparticulate, Nanoporous Ceramics Made under Ambient Conditions, *Microscopy and Microanalysis '04: Proc. 62nd Annual Meeting of Microscopy Society of America*, **10**, 404-05, (2004).

GEOPOLYMERIZATION OF RED MUD AND RICE HUSK ASH AND POTENTIALS OF THE RESULTING GEOPOLYMERIC PRODUCTS FOR CIVIL INFRASTRUCTURE APPLICATIONS

Jian He (Corresponding Author)
Graduate Research Assistant
Dept. of Civil and Environmental Engineering, Louisiana State University
Baton Rouge, LA 70803, USA

Guoping Zhang
Assistant Professor
Dept. of Civil and Environmental Engineering, Louisiana State University
Baton Rouge, LA 70803, USA

ABSTRACT

This work presents the results of a study on reusing red mud (RM), an abundant alumina refinery waste produced by the Bayer process, and rice husk ash (RHA), a major waste from combustion of rice husk, as raw materials for the production of geopolymers that are environmentally friendly and only require low energy to make and have diverse potential applications. A wide range of parameters in the geopolymerization reaction, consisting of RHA to RM weight ratio, different particle size of RHA, and variable concentrations of sodium hydroxide solution, were examined to understand their influence on the compressive strength of the end products – RM&RHA-based geopolymers. The composition of RM, RHA, and RM&RHA-based geopolymers was characterized by X-ray diffraction. Moreover, the results of unconfined compression testing indicate that the compressive strength of the studied RM&RHA-based geopolymers is in the range of 3.2 to 20.5 MPa, which is comparable with that of almost all Portland cements. In addition, the utilization of RM&RHA-based geopolymers in practice is able to bring both environmental and economic advantages. The findings suggest that these two plentiful wastes, RM and RHA, can be reused to make geopolymers that can find applications in civil infrastructure constructions.

INTRODUCTION

Red mud (RM) is the major industrial waste produced by the Bayer process for the extraction of alumina from bauxite ores,[1] which is characterized by strong alkalinity and high water content.[2] In general, RM includes mainly iron oxide (mostly hematite), alumina, and some heavy metals. Depending on the quality and purity of the bauxite ore, the quantity of RM generated varies from 55-65% of the processed bauxite.[3] According to recent US Geological Survey reports,[4,5] bauxite ore mined globally amounts to 202 million tons (MT) in 2007, 205 MT in 2008, and 201 MT in 2009. Hence it is estimated that the worldwide production of RM is approximately 120 MT annually. In fact, utilization of RM is a first-priority issue for all alumina plants.[6] In the past, RM was disposed of either directly into the sea or onto the land in waste ponds. At present, the way of sea disposal has been prohibited all over the world due to stricter environmental constraints.[6] Although extensive research on RM utilization has been conducted in the past decades,[7-12] a widely accepted technology that can be employed for the recycle of RM is still not available at present. Thus, the way of dumping it in the waste ponds is mainly used by alumina plants currently. According to Hungarian Ministry of Foreign Affairs (2010),[13] a RM spill occurred in Hungary on October 4, 2010 due to the levee breach of the giant RM pond. This tragedy killed at least 8 people, seriously harmed hundreds of residents and the environment, which was the worst ecological disaster in Hungary. Figure 1 illustrates a contaminated village in this RM spill. The disaster demonstrates that the current RM disposal method that is widely used globally is not reliable and safe. Thus, new trustworthy and environmentally-friendly disposal methods are urgently needed.

Figure 1. A contaminated village in the RM spill in Hungary [13]

Rice is one of the major crops grown throughout the world,[14] sharing equal importance with wheat as the principal staple food and a provider of nourishment for the human beings around the world.[15] Rice milling generates a by-product known as husk, which is the hard protecting covering of grains of rice. Finally, rice husk ash (RHA) is obtained by burning the rice husk in the boiler, which is collected from the particulate collection equipment attached upstream to the stack of rice-fired boilers.[16] Per a RHA market study in 2003,[17] approximately 600 MT of rice paddy is produced globally each year. On average 20% of the rice paddy is husk, giving an annual total production of 120 MT. Juliano (1985) pointed out that rice husk is one of the most intractable agricultural wastes since its tough, woody, and abrasive nature along with high silica content has made disposal very difficult.[18] Moreover, rice husk is unusually high in ash compared to other biomass fuels, about 20%, which is influenced by the variety, climatic conditions, and geographical location.[19-20] Thus, it is estimated that about 24 MT of RHA is formed worldwide every year. RHA is highly porous and lightweight with a very high external surface area and contains amorphous silica in high content (usually 90 – 95 wt.%). In addition, Costa et al. (1999) stated that the most common method of disposal of RHA is dumping on waste land, thus creating an environmental hazard through pollution and land dereliction problems.[21] Therefore, effective and environmentally-friendly ways of disposal of RHA are also urgently needed due to the immerse amount of RHA generated annually.

Nowadays, geopolymerization technology has attracted more and more attention as a solution of utilizing solid waste and by-products, which provides a mature and cost-effective solution to many problems where hazardous residue has to be treated and stored under critical environmental conditions.[22] Geopolymerization is to react silica-rich and alumina-rich solids with a high alkaline solution to form amorphous to semi-crystalline aluminosilicate polymers, which exhibit excellent physical and chemical properties and hence have a series of applications in many fields such as insulation materials, construction materials, and waste containment. In general, any materials that contain mostly silica (SiO_2) and alumina (Al_2O_3) in amorphous form are a possible source for the production of geopolymer.[21] In fact, geopolymerization of RM and RHA takes the advantage of their characteristics: strong alkalinity (NaOH) and amorphous alumina (Al_2O_3) in RM and the presence of a great amount of amorphous silica (SiO_2) in RHA. This work was to investigate the potential utilization of RM and RHA as raw materials for the production of geopolymers that can be used in civil infrastructure constructions.

MATERIALS AND METHODS

Materials

The raw materials used for geopolymer synthesis include RM slurry (Gramercy Alumina, LLC, USA), RHA (Agrilectric Power, Inc, USA), sodium hydroxide (purity quotient: 99%, Sigma-Aldrich Co., USA), and deionized water. The RM slurry has a pH 11.9 and contains mainly water with dissolved Na-aluminate ($NaAlO_2$) and sodium hydroxide ($NaOH$) as liquid phase, and hematite (Fe_2O_3) and alumina (Al_2O_3) as solid phase (Table I). The slurry was air-dried, homogenized, and pulverized to a powder until all solids passed a #60-mesh sieve, in order to facilitate geopolymerization and minimize the influence of compositional variation on the geopolymers. The RHA mainly contains silica, potassium oxide, and carbon (Table I).

TABLE I. Chemical composition and concentrations (wt.%) of RM and RHA

Material	Red mud	RHA
SiO_2	1.2	91.5
Al_2O_3	14.0	-
Fe_2O_3	30.9	-
$NaOH$	20.2	-
$NaAlO_2$	23.0	-
CaO	2.5	-
MgO	-	-
S	-	-
K_2O	-	2.3
TiO_2	4.5	-
MnO	1.7	-
C	-	6
Total	98.0	99.8

Methods

Geopolymer synthesis started with dry mixing the powders at a selected RHA to RM weight ratio (RHA/RM), followed by adding sodium hydroxide solution with different concentrations to the powder mixture (containing both RM and RHA) at a solution/solid = 1.2. The mixture was then stirred for > 15 minutes to ensure sufficient reaction (e.g., dissolution) between the powder and solution, resulting in the formation of geopolymer precursor paste. The precursor was then poured into cylindrical molds with an inner diameter of 2 cm and height of 5 cm (i.e., an aspect ratio of 2.5 to minimize the end effects), followed by curing in an ambient environment (e.g., room temperature) for 14 days (Although high curing temperature can enhance the properties of final geopolymeric products, it is difficult to apply in practice). The specimens were then demolded, followed by prolonged curing in exposed conditions. In this study, RHA was used as an alternative of sodium silicate that was usually used when synthesizing geopolymers since it contains plenty of amorphous SiO_2 (Table 1). In this work, general geopolymer samples were made with parameters of RHA/RM = 0.4, regular size of RHA, and 4 M sodium hydroxide solution. Samples with different parameters were also synthesized to investigate their influence on the compressive strength of the resulting geopolymeric products. In addition, unconfined compression tests were performed on cured cylindrical specimens using an automated GeoTAC loading frame (Trautwein Soil Testing Equipment, Inc., USA) at a constant strain rate of 0.5%.

RESULTS

Compressive Strength

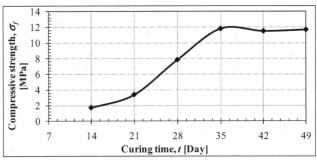

Figure 2. Influence of curing time on the compressive strength of RM&RHA-based geopolymers.

Figure 2 plots the compressive strength (σ_f) against curing time for RM&RHA-based geopolymer samples with aforementioned parameters (e.g., RHA/RM = 0.4, regular size of RHA, and 4 M sodium hydroxide solution). It is clear that σ_f increases initially with curing time, and then reaches a constant value for even prolonged curing. The strength stabilizes at 11.9 MPa after 35 days curing, suggesting that the RM&RHA-based geopolymers can achieve complete curing in 35 days. To ensure complete curing and minimize the undesired influence of incomplete curing on the compressive strength, all geopolymer samples discussed in the following sections were cured for 60 days.

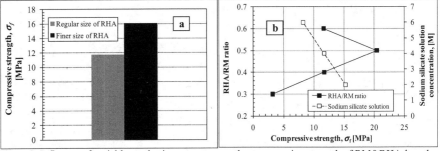

Figure 3. Influence of variable synthesis parameters on the compressive strength of RM&RHA-based geopolymers: (a) RHA particle size; (b) RHA/RM ratio and sodium hydroxide solution concentration.

In order to understand the factors affecting the compressive strength of RM&RHA-based geopolymers, Figure 3a and 3b illustrate the influence of RHA particle size and the influence of RHA/RM ratio and sodium hydroxide solution concentration, respectively. To examine the influence of one specific parameter, only the studied parameter was changed while other parameters remained unchanged. As seen in Figure 3a, geopolymer samples synthesized with finer size of RHA has higher σ_f (16.1 MPa) than the ones made with regular size of RHA (11.9 MPa), suggesting that finer RHA particle size causes positive effect on the geopolymerization of RHA and RM. As seen in Figure 3b, the compressive strength of geopolymer samples increases with RHA/RM ratio (i.e., from 0.3 to 0.5) initially, then decreases with it (i.e., from 0.5 to 0.6), indicating the optimum RHA/RM ratio for the synthesis is 0.5. Moreover, σ_f of geopolymer samples decreases with the increase of concentration of sodium hydroxide solution at a linear relationship.

XRD Analysis

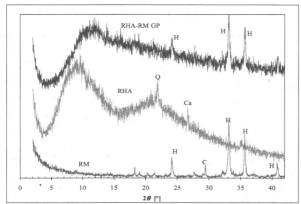

Figure 4. XRD patterns of RM, RHA, and RHA-RM GP (H= hematite, C = calcite, Q = quartz, Ca = carbon).

Figure 4 displays the XRD patterns of RM, RHA, and cured RM&RHA-based geopolymers. RM shows some unidentified peaks and a few sharp peaks that are mainly from hematite and calcite, but no observable broad humps, suggesting that the amorphous phases are not present at large quantity. By comparison with its chemical composition (Table 1), alumina mainly presents as amorphous phases. Thus, red mud provides mainly Al (in the form of amorphous Al_2O_3 or dissolved $NaAlO_2$) and NaOH but little Si to geopolymerization. RHA shows 2 huge and broad humps at 5-15 and 15-30 $°2\theta$, respectively, indicating the presence of amorphous phases. A few sharp peaks also indicate the presence of crystalline phases such as quartz and carbon. By comparison with its chemical composition (Table I), silica in RHA are present as both amorphous and crystalline (quartz) phases. Furthermore, the overwhelming majority of silica is in amorphous form, which agrees with the previous studies.[23-25] The pattern of RM&RHA-based geopolymers shows a huge, broad, and nonsymmetrical hump between 5 – 40 $°2\theta$ with a few sharp peaks from crystalline phases, suggesting that amorphous phases are present in RM&RHA-based geopolymers, which coincides with a currently common agreement that aluminosilicate polymers are amorphous to semi-crystalline.[26-28] The presence of sharp peaks of crystalline phases from parent material confirms that the crystalline phases are not reactive or involved in geopolymerization, but simply present as inactive fillers in geopolymer network.[1]

DISCUSSION

There are actually three parent materials involved in the synthesis of RM&RHA-based geopolymers: RM, RHA, and sodium hydroxide solution. As discussed above, only amorphous phases in raw materials participate in geopolymerization reaction. Among the three raw materials, RM provides NaOH, Al_2O_3, and $NaAlO_2$; RHA provides amorphous SiO_2; sodium hydroxide solution provides NaOH. According to the XRD results, the final geopolymeric products are not pure geopolymer matrix, but the composites consisting of neoformed geopolymer structure and crystalline phases from parent materials, which is consistent with previous study.[1]

The compressive strength (σ_f) of the studied RM&RHA-based geopolymers ranges from 3.2 to 20.5 MPa, comparable to that of all Portland cement (9 – 20.7 MPa) except Type III (24.1 MPa) Portland cement.[29] As such, RM&RHA-based geopolymers can be used as a cementitious material to replace Portland cement in certain engineering applications, such as civil infrastructure constructions, which also brings both environmental and economic advantages. First, two of the three raw materials, RM and RHA, used in the synthesis of RM&RHA-based geopolymers are abundant industrial and agricultural wastes. Geopolymerization technology may provide an effective and environmentally-friendly way to disposal of them. Thus, geopolymerization of RM and RHA can save not only the expenses for waste disposal and long-term monitoring and maintenance of waste containment facilities, but also the costs for manufacturing of Portland cement. Second, recycle of the two abundant wastes can minimize the potential damage of the waste to the environment and human health. Third, the hematite in the geopolymer network is actually highly absorptive for heavy metals, which can act as a reactive barrier to filter the contaminants transported through percolating inside the geopolymer matrix. Finally, the elimination of Portland cement usage can save the energy associated with cement production and reduce the CO_2 emission caused by firing carbonates.[26] Based on the previous literature,[22,30] the geopolymer composites have the ability to immobilize toxic chemical and radioactive waste within their own structures. Thus, RM&RHA-based geopolymers could also be used in other potential applications such as waste containment.

As the results suggested, complete curing duration of RM&RHA-based geopolymers is 35 days, much longer than that of geopolymers synthesized with other parent materials such as metakaolin or fly ash, which is sometimes too long for certain applications. Therefore, further research is desired to investigate how to accelerate the curing time of RM&RHA-based geopolymers, which will definitely improve the feasibility of application of RM&RHA-based geopolymers in practice. Actually, the prior research has pointed out that geopolymers can exhibit a wide variety of properties and characteristics depending on the raw material selection and processing conditions and addition of fly ash can make the geopolymer more durable and stronger.[2] Additionally, the properties of RHA are firmly associated with variety, climatic conditions, and geographical location of rice paddy and combustion conditions such as burning duration and burning temperature.[24-25] Therefore, addition of fly ash and using different type of RHA are two superior points to investigate in the further research to have a better understanding of factors affecting the properties and characteristics of RM&RHA-based geopolymers so that the ones with improved properties can be produced.

CONCLUSIONS

This paper presents an experimental study that investigates the potential reuse of RM, an industrial waste from alumina refining, and RHA, an agricultural waste from combustion of rice husk, via geopolymerization reactions with sodium hydroxide solution, the only non-waste material. A wide variety of parameters involved in the synthesis, including RHA/RM ratio, RHA particle size, and concentration of sodium hydroxide solution, were examined to understand their influence on the compressive strength of the RM&RHA-based geopolymers. The results indicate that the σ_f of the final products vary significantly with different synthesis conditions. RM&RHA-based geopolymers synthesized with finer size of RHA, RHA/RM = 0.5, and 2M sodium hydroxide solution should have the best mechanical performance (i.e., highest σ_f). For the studied RM&RHA-based geopolymers, they exhibit a compressive strength of 3.2 - 20.5 MPa, comparable with that of all Portland cement (9 – 20.7 MPa) except Type III (24.1 MPa) Portland cement. Therefore, this study demonstrates that both RM and RHA can be reused via geopolymerization and the end products, RM&RHA-based geopolymers, are a viable, promising cementitious material with potential utilization in civil infrastructure constructions.

ACKNOWLEDGMENTS

This study was supported by the fund awarded to G.Z. J.H. also received partial support, the Graduate Enhancement Award, from the LSU Graduate School. RM samples were provided by Gramercy Alumina, LLC. RHA samples were provided by Agrilectric Research Company, Inc.

REFERENCES
1. G. Zhang, J. He, and R. P. Gambrell, Synthesis, Characterization, and Mechanical Properties of Red Mud-Based Geopolymers, *Transportation Research Record: Journal of the Transportation Research Board*, **2167**, 1-9 (2010).
2. W. Cundi, Y. Hirano, T. Terai, R. Vallepu, A. Mikuni, and K. Ikeda, Preparation of Geopolymeric Monoliths from Red Mud-PFBC Ash Fillers at Ambient Temperature, *Proceedings of the World Congress Geopolymer 2005,* Saint-Quentin, France, 85-87 (2005).
3. R. K. Paramguru, P. C. Rath, and V. N. Misra, Trends in Red Mud Utilization – A Review, *Mineral Processing & Extractive Metallurgy*, **26**, 1-29 (2005).
4. Mineral Commodity Summaries. *US Geological Survey*, January 2009. http://minerals.usgs.gov/minerals/pubs/mcs/2009/mcs2009.pdf. Accessed Dec. 5, 2010.
5. Mineral Commodity Summaries. *US Geological Survey*, January 2010. http://minerals.usgs.gov/minerals/pubs/mcs/2010/mcs2010.pdf. Accessed Dec. 5, 2010.
6. D. D. Dimas, I. P. Giannopoulou, and D. Panias, Utilization of Alumina Red Mud for Synthesis of Inorganic Polymeric Materials, *Mineral Processing & Extractive Metallurgy Review*, **30**, 211-39 (2009).
7. J. I. Glanville and P. E. Winnipeg, *Bauxite Waste Bricks-International Development Research Center*, (1991). http://www.idrc.ca/library/document/099941. Accessed Dec. 5, 2010.
8. M. S. N. Singh and P. M. Prasad, Preparation of Special Cements from Red Mud, *Waste Management*, **16**, 665-70 (1996).
9. M. S. N. Singh, S. N. Upadhayay, and P. M. Prasad, Preparation of Iron Rich Cements Using Red Mud, *Cement and Concrete Research*, **27**, 1037-46 (1997).
10. A. M. Marabini, P. Plescia, D. Maccari, F. Burragato, and M. Pelino, New Materials from Industrial and Mining Wastes: Glass-ceramics and Glass- and Rock- Wool Fiber, *International Journal of Mineral Processing*, **53**, 121-34 (1998).
11. N. Yalcin and V. Sevinc, Utilization of Bauxite Waste in Ceramic Glazes, *Ceramics International*, **26**, 485-93 (2000).
12. R. U. Ayres, J. Holmberg, and B. Andersson, Materials and the Global Environment: Waste Mining in the 21st Century, *Materials Research Society Bulletin*, 477-80 (2001).
13. Hungary Ministry of Foreign Affairs, Information on the Accidental Pollution and Related Mitigation Meansures of the Red Mud Spill at Ajka, http://www.mfa.gov.hu/kulkepviselet/CZ/en/en_Hirek/V%C3%B6r%C3%B6siszap_en.htm, accessed by Dec. 15, 2010.
14. D. N. Mbui, P. M. Shiundu, R. M. Ndonye, and G. N. Kamau, Adsorption and Detection of Some Phenolic Compounds BY Rice Husk Ash of Kenyan Origin, *Journal of Environmental Monitoring*, **4**, 978-84 (2002).
15. L. Z. Gao, Population Structure and Conservation Genetics of Wild Rice Oryze Ruflpogon (Poaceae): A Region-Wide Perspective from Microsatellite Variation, *Molecular Ecology*, **13**, 1009-24 (2004).

16. T. K. Naiya, A. K. Bhattacharya, S. Mandal, and S. K. Das, The Sorption of Lead (II) Ions on Rice Husk Ash, *Journal of Hazardous Materials*, **163**, 1254-1264 (2009).
17. Rice husk ash market study. Bronzeoak Ltd, 2003, http://webarchive.nationalarchives.gov.uk/+/http://www.berr.gov.uk/files/file15138.pdf
18. B. O. Juliano, In Rice: Chemistry and Technology, 2nd ed., *American Association of Cereal Chemists*, St. Paul, MN, USA, 1985.
19. S. Asavapisit and N. Ruengrit. The Role of RHA Blended Cement in Stabilizing Metal Containing Wastes, *Cement and Concrete Composites*, **27**, 782-87 (2005).
20. E. A. Basha, R. Hashim, H. B. Mahmud, and A. S. Muntohar, Stabilization of Residual Soil with Rice Husk Ash and Cement, *Construction and Building Materials*, **19**, 448-53 (2005).
21. H. M. Costa, L. L. Y. Visconte, R. C. R. Nunes, and C. R. G. Furtado, The Effect of Coupling Agent and Chemical Treatment on Rice Husk Ash – Filled Natural Rubber Composites, *Journal of Applied Polymer Science*, **76**, 1019-27 (1999).
22. R. D. Hart, J. L. Lowe, D. C. Southam, D. S. Perera, P. Walls, E. R. Vance, T. Gourley, and K. Wright, Aluminosilicate Inorganic Polymers from Waste Materials, *Green Proceeding Conference*, Newcastle, 2006.
23. K. Y. Foo and B. H. Hameed, Utilization of Rice Husk Ash as Novel Adsorbent: A Judicious Recycling of the Colloidal Agricultural Waste, *Advances in Colloid and Interface Science*, **152**, 39-47 (2009).
24. A. Muthadhi, R. Anitha, and S. Kothandaraman, Rice Husk Ash – Properties and Its Uses: A Review, *Journal of the Institution of Engineers*, **88**, 50-6 (2007).
25. D. S. Chaudhary, M. C. Jollands, and F. Cser, Recycling Rice Hull Ash: A Filler Material for Polymeric Composites, *Advances in Polymer Technology*, **23**, 147-155 (2004).
26. J. Davidovits, Geopolymers: Inorganic Polymeric New Materials, *J. Therm. Anal.*, **37**, 1633–56 (1991).
27. P. Duxson, A. Fernandez-Jimenez, J. L. Provis,, G. C. Lukey, A. Palomo, and J. S. J. Van Deventer, Geopolymer Technology: The Current State of the Art, *J. Mater. Sci.*, **42**, 2917–2933 (2007).
28. J. He, G. Zhang, S. Hou, and C. Cai, Geopolymer-Based Smart Adhesives for Infrastructure Health Monitoring: Concept and Feasibility, *Journal of Materials in Civil Engineering (ASCE)*. (In press).
29. ASTM – American Society for Testing and Materials. *Annual Book of ASTM Standards (C 150 – 05)*, 2005.
30. J. G. S. Van Jaarsveld, J. G. S. Van Deventer, and L. Lorenzen, The Potential Use of Geopolymeric Materials to Immobilize Toxic Metals: Part I. Theory and Applications, *Mineral Engineering*, **10**, 659-669 (1996).

THE EFFECT OF ADDITION OF POZZOLANIC TUFF ON GEOPOLYMERS

Hani Khoury*, Islam Al Dabsheh, Faten Slaty, Yousef Abu Salha,
University of Jordan, Amman, 11942, Jordan

Hubert Rahier
Research Group of Physical Chemistry and Polymer Science (FYSC), Vrije Universiteit Brussel
(VUB)

Muayad.Esaifan, Jan Wastiels
Department of Mechanics of Materials and Constructions (MEMC), Vrije Universiteit Brussel (VUB)

ABSTRACT
    A geopolymerization process was developed by activating local raw kaolinite with quartz and tuff
fillers and alkaline sodium hydroxide solution. The new geopolymers were cured at 80 °C for 24 hr.
The effect of addition of pozzolanic tuff on the physical and chemical properties of the geopolymers
was investigated. The results have indicated that the replacement of 50% sand by pozzolanic tuff lead
to an increase in compressive strength after 14 days curing time under dry and wet conditions. The
geopolymers' mechanical strength increase upon heat treatment till 500 °C before it drops down to 14
MPa at 900 °C. The mechanical performance remarkably increases after a heat treatment above 900 °C
and reaches its maximum (68.3 MPa) at 1100 °C before the collapse of the geopolymer as a result of
melting in the furnace at 1200°C and vitrification/crystallization upon cooling. Crystalline sodium
and/or calcium aluminum silicate phases are responsible for the high compressive strength. Heating the
geopolymer at 1100°C for 24 h did not show a collapse of the texture and this is an important
indication of refractoriness.
    The high strength, heat resistance, low production cost, low energy consumption, and low CO_2
emissions suit the use of the geopolymers as a green construction material.

INTRODUCTION
Geopolymer technology has recently attracted researchers because the products are non-combustible,
heat-resistant, formed at low temperatures, fire resistant and have environment friendly applications [1;
2; 3 and 4]. Such products could be useful as construction materials [5 and 6]. Geopolymers have been
proposed as an alternative to traditional Ordinary Portland Cement (OPC) for use in construction
applications, due to their excellent mechanical properties [7]. Their physical behavior exceeded that of
Portland cement in respect of compressive strength, resistance to fire, heat and acidity, and as a
medium for the encapsulation of hazardous or low/intermediate level radioactive waste [4; 8; 9 and 10].
    Stabilization techniques of clay material include mechanical, physical, and chemical stabilization.
Physical stabilization utilizes compaction on its texture that changes its density, mechanical strength,
compressibility, permeability and porosity.
Chemical stabilization changes the properties of the material through adding fillers or chemicals.
Physico-chemical reactions take place between the components and/or new phases that bind or coat the
components. Kaolinite could be hardened and transformed into aluminosilicate polymers through
chemical polymerization reactions.
    Clay minerals (calcined clay), mining wastes, and slag are considered as a good source of
aluminosilicate precursor. Geopolymers consist of three-dimensional mineral phases resulting from
the polymerization of two dimensional sheet-like aluminosilicates in an alkaline solution. The exact
mechanism of geopolymerization is not known precisely until now. The structure maintains electrical

neutrality as a result of aluminum substitution for silicon in the tetrahedral layer and the compensation of the negative charge by the available cations such as Na^+ [1; 2; 11 and 12].

Inexpensive functional fillers like silica sand and zeolitic tuff were used in different proportions to help in stabilizing the produced geopolymers [6]. The excellent mechanical properties of the geopolymers have attracted the researchers to focus on the effect and application of different raw materials on the compressive strength, chemical impurities, and the effect of the chemical composition of the alkali activating solutions [13]. New low temperature functional geopolymeric materials were prepared exhibiting adsorption capacity for pollutants [8 and 16].

In this work the kaolinite will not be calcined before the geopolymerization. This even further reduces the greenhouse gas emissions.

The following work will focus on evaluating the influence of adding volcanic (pozzolanic) tuff as filler to the kaolinitic mixture to improve its mechanical properties and to evaluate its effect on efflorescence. The mechanical performance, stability, thermal behavior (up to 1200°C) and phase transformation of the low temperature geopolymers will be evaluated.

MATERIALS AND TECHNIQUES

Materials:

Jordanian kaolinitic clay (as a source of aluminum silicate) from El-Hiswa deposit with a purity of 60% was used. The kaolinitic clay deposits are located in the south of Jordan about 45 km to the east of Al-Quweira town [8; 17]. Preparation of the Jordanian kaolinite samples involved crushing, grinding and sieving of an oven dried sample (at 105 °C) to obtain a grain size less than 425 μm. Then the samples were mixed in a 50 L plastic drum for several times for homogenization. The plasticity limit of El-Hiswa kaolinite was measured according to the ASTM D4318 [9] and was found to be 22%.

Silica sand with 99% quartz content was used as a filler to provide high mechanical properties. The sand was washed with distilled water and sieved to obtain a grain size between 100 and 400 μm.

Volcanic tuff (pozzolana) from north east Jordan was used as filler to substitute for quartz in the mixture to minimize the efflorescence effect [4; 10]. Figure shows the size range of the powdered tuff between 40 nm and 25 μm.

Figure 1. SEM photomicrograph of the particle size of the powdered tuff

NaOH (Grade 96 %) was used as an alkaline activator for the dissolution of the aluminoslicate precursor. Water was the reaction medium and the optimum water content was close to the plasticity limit. The optimum curing time at 80°C was determined to be around 24 hours [4, 14]. The optimal ratios of the mixture were determined depending on the best compressive strength, the optimal curing temperature and time for the geopolymer specimens [4, 14]. The composition of the optimized mixture to produce geopolymers is given in Table 1. In the text the samples will be named according to the % of tuff as filler, thus 50% tuff is sample S3.

Table 1. Composition in grams of the geopolymer mixtures (cured at 80 °C for 1 day).

Series	Clay (Raw Kaolinite)	Sand	Pozzolanic Tuff	NaOH (Solid purity 96%)	Distilled Water
S1	100	100	0	16	22
S2	100	75	25	16	22
S3	100	50	50	16	22
S4	100	25	75	16	22
S5	100	0	100	16	22

Fabrication of geopolymer specimens:

Homogeneous mixtures were prepared (Table 1) using a controlled speed mixer (mixing speed was 107 rpm for 2 min followed by 198 rpm for 10min). Good mixing is important to obtain homogeneous and comparable specimens and to avoid the agglomeration of the mixture. Each mixture (series) was divided into specimens (50 g each). The mixture was molded immediately after weighing to avoid drying and decrease of the workability of the mixture. The paste was molded in a stainless steel cylinder (diameter of 25mm and height of 45mm) at a pressure of about 15 MPa (Carver hydraulic laboratory press). The molded specimens of each series were cured by placing them uncovered in a ventilated oven (Binder-ED115) at 80 °C for 24 hr. The compressive strength for the five geopolymer series was measured after 1, 7, 14 and 28 days using CONTROLS testing machine (Model T106 modified to suit with standard testing). The displacement rate was 2 mm/min. The geopolymer with the composition 50 % tuff (series 3) gave the highest compressive strength and was chosen to study the thermal effect.

After curing, the specimens were cooled down and stored (aged) at room temperature. Wetting-drying test was used to determine the resistance of specimens to repeated wetting and drying cycles. Wet samples were aged for 1, 7, 14, and 28 days. The specimens were submerged in distilled water at room temperature and then dried in an oven at 40 °C for the next 24h.

The specimens were tested for physical properties. The density, water absorption and electrical conductivity (Ec) of the immersed specimens were measured [4, 14]. The electrical conductivity is measured on the immersion solution (three samples/point). The Electrical conductivity (EC) was measured (using WTW multi-line P4, pH and EC meter) as a function of time; until they become constant (complete leaching of water-soluble salts is achieved).

Analytical techniques:

The zero measurements (reference) were recorded immediately after curing at 80°C for 24 hours (1 day). For density measurements, the specimens were weighed using an electronic balance (SPE2001, Scout Pro) and their dimensions were measured using a digital micrometer (electronic digital caliber, 0.155mm).The plasticity limits were determined according to the ASTM D4318 [15].

The effect of adding pozzolanic tuff on the mineral phases of the geopolymers (the five series) was studied by using X-ray diffraction techniques. Random preparations (packed powder with no orientation) were made to identify the crystalline phases of the materials. Representative portions of the ground heated geopolymers were X-rayed using Philips 2KW model, Cu Kα radiation (λ= 1.5418 Å) with a scan rate of 2°/min. X-ray diffractograms were recorded for the five mixtures to identify the phase changes.

The geopolymer specimens (50 % tuff) were heated in the furnace up to 100, 200, 300, 400, 500, 600, 700, 800, 900, 1000, 1100 and 1200 °C (25°C /minute). After heating for 24 hours (soak time), the specimens were brought back to ambient temperature before further testing. The compressive strength for the fabricated heated geopolymer specimens (series 3) up to 1200 °C was measured using the above mentioned testing machine.

The phase changes of the heated geopolymers (50 % tuff, series 3) were studied by using the above described X-ray diffraction techniques. X-ray diffractograms were recorded for powdered geopolymers at 80 up to 1200 °C to detect the phase changes.

SEM/EDX was also used for obtaining mineralogical, chemical and textural details for the five geopolymer mixtures and the heated geopolymer specimens after 24 hours. The platinum coated geopolymer samples were scanned using high-energy beam of primary electrons in a raster scan pattern using model FEI- INSPECT-F50 of SEM/EDX.

RESULTS AND DISCUSSION

The chemical composition of the added pozzolanic tuff is given in Table 2. The composition is equivalent to basalt (18). The alkali content is about 4.5 % and CaO is about 8.7 %. The tuff sample was obtained from Al Rajhi Cement Factory and is used in industry as additive to produce pozzolanic Portland cement.

Figure 2 shows the relationship between the plastic limit and the tuff %. The plasticity limits are important to optimize the amount of water needed for the specimens' preparation. The amount of distilled water was determined according to the plasticity limit for the mixture of kaolinite, sand and tuff. The figure indicates that the higher the amount of tuff the lower is the plastic limit is this logic since the tuff, like sand does not show plastic properties.

Table 2: Chemical composition of the pozzolanic tuff (XRF results of Al Rajhi Cement Factory).

Oxide	CaO	SiO_2	Al_2O_3	Fe_2O_3	MgO	SO_3	K_2O	Na_2O_3	TiO_2
%	8.74	44.63	14.17	13.21	9.48	0.09	1.41	3.12	2.42

Figure 3 illustrates the effect of addition of tuff as a filler to substitute for quartz in the mixture on the compressive strength. The replacement of 25 to 75% sand by pozzolanic tuff as filler has in general improved the compressive strength of the samples after 1, 7, 14, and 28 days. The compressive strength is however again on the initial level with the 100% substitution of sand by tuff. Remark that after one day the strength of the 100% tuff sample is clearly lower than the 100% sand sample, probably indicating a less availability of dissolved silica (100% tuff means no quartz as filler). In general, as indicated in figures 3 and 10, the presence of pozzolanic tuff as a substitute of quartz indicates higher reactivity causing rapid setting when cured at 80 °C, leaving behind undissolved tuff (anorthite and diopside). The compressive strength of the immersed samples shows the same trend namely the series with 25- 75% tuff have an increased compressive strength. Ageing has improved the compressive strength gradually.

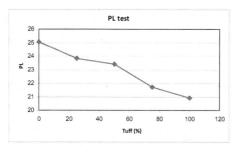

Figure 2. The relationship between plastic limit and tuff %. The line is drawn as a guide to the eye.

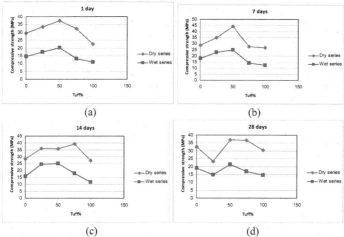

(a)   (b)

(c)   (d)

Figure 3 (a-d). The effect of pozzolanic tuff and ageing (1, 7, 14, and 28 days at room temperature) on the compressive strength.

Figure 4 illustrates that the use of 100% tuff as filler improves the compressive strength (from 25 to 30 MPa) with ageing (after 28 days). It indicates a possible further chemical reaction and crystallization of the amorphous phases that act as a binding material. The same trend appears also with the immersed specimens.

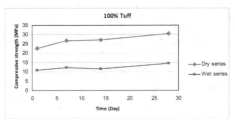

Figure 4. The effect of ageing on compressive strength

The electrical conductivity of the water in which the geopolymer specimens were immersed was measured. The results are plotted in Figure 5. The plot illustrates also the effect of ageing after 1, 7, 14 and 28 days. The figure indicates that the higher the tuff %, the higher is the Ec values as a result of the release of dissolved alkalis. The increased electrical conductivity of the immersed samples could be related to water-soluble salts, that cause the efflorescence problem. Efflorescence is the result of carbonation of residual NaOH, since the end geopolymer products contain sodium carbonate as major water-soluble salt.

A not expected result is that the electrical conductivity increases with ageing time of the specimens. The lowest solubility is measured after one day ageing and for longer times of ageing and thus less reactivity and higher electrical conductivity could be related to the increase of quartz (Figure 5). In the presence of quartz as a filler (less tuff), less reactivity, more dissolved solids, and higher Ec is indicated. This means that more material can be dissolved in the water after longer reaction time (ageing). No explanation was found for this particular behavior and further investigation is ongoing.

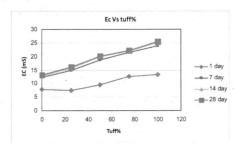

Figure 5. The effect of tuff % and ageing on electrical conductivity

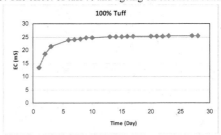

Figure 6. The effect of ageing on electrical conductivity (adding 100% tuff)

The effect of tuff % on the density of the geopolymers is illustrated in Figure 7. The density is at maximum as a result of adding 25 and 50% tuff. The figure indicates that the maximum reaction with tuff takes place by adding 50 % tuff. The higher is the reactivity with tuff; the lower is the porosity (higher density) as a result of polymerization. The same trend appears with the immersed specimens,. The addition of 25% and 50 % tuff produces higher density geopolymers than other concentration.

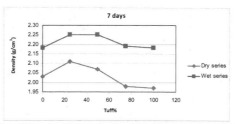

Figure 7. The effect of tuff % on the density of the geopolymers after 7 days

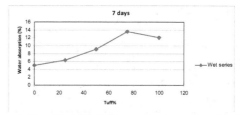

Figure 8. The effect of tuff % on water absorption after 7 days

The effect of tuff % on water absorption is illustrated in Figure 8 and Figure 9. Water absorption increases with the increase of tuff % after 7 days Figure 8.
The maximum water absorption with the increase of tuff % is due to the presence of voids and a smaller amount of crystalline fraction in the matrix of the geopolymers as will be shown later.

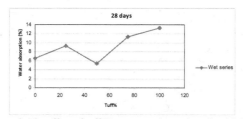

Figure 9. The effect of tuff % on water absorption after 28 days

The X-ray diffraction results of the produced geopolymers from the different series (Table 1) are illustrated in Figure 10. The X-ray diffraction patterns of the five geopolymer mixtures are compared with the added kaolinite and tuff to indicate any phase changes.

The most important mineral phase of the produced geopolymer from series 1 to 4 (up to 75% tuff) at 80°C is sodalite. The same results were obtained by other researchers working on Jordanian kaolinite [4; 10]. As identified by XRD, the mineral phases of the geopolymer cured at 80 °C (series 1, 0 % tuff) are mainly composed of kaolinite, muscovite-illite, quartz and Na-Al silicate phases (sodalite). Sodalite as the new geopolymer phase stays in all the mixtures (0 -100 % tuff). Kaolinite peaks became weaker in the different series as a result of the chemical attack during the geopolymerization. The X-ray diffraction results (Figure **10**) indicate the presence of residual kaolinite as a result of incomplete reaction with Na (OH). The diffraction peaks of the mineral phases in the tuff mostly disappeared. The alkaline solution reacts with tuff to produce amorphous and crystalline phases of Na-Ca-Al silicates similar to phillipsite and nepheline.

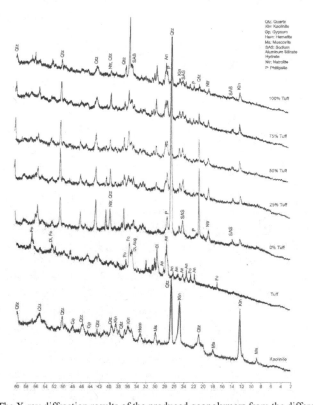

Figure 10. The X-ray diffraction results of the produced geopolymers from the different series.

Figure 11 to Figure 15 illustrate selected SEM photomicrographs of the different series cured at 80 °C after 24 hours. The figures indicate the presence of residual kaolinite in the different series. Residual kaolinite shows reaction rims, embayment and etching. New Na-Al-silicate phases as indicated from EDX data are embedded in the matrix (Figure 11).

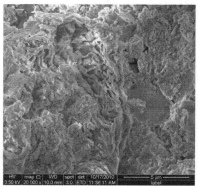

Figure 11. SEM photomicrograph of the first series (0% tuff).

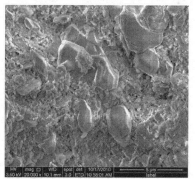

Figure 12. SEM photomicrograph of the second series (25 % tuff).

Figure 12 illustrates also the new phases in the matrix after adding 25 % tuff. The figure illustrates the formation of idiomorphic quartz. Crystalline and amorphous sodium potassium calcium aluminum silicate phases are also retrieved in the matrix.

Figure 13 illustrates the SEM photomicrograph after the addition of 50% tuff. The figure shows geopolymerized crystalline and amorphous matrix.

Figure 13. SEM photomicrograph of octahedral zeolite phases growing at the expense of amorphous polymerized phases (50% tuff).

Figure 14 illustrates the formation of new amorphous and crystalline phases in the matrix at the expense of kaolinite and tuff. The addition of 75 % tuff indicates higher solubility of the tuff and the formation of amorphous material.

Figure 14. SEM photomicrograph of amorphous and crystalline polymerized phases (75% tuff).

Figure 15 illustrates the formation of new amorphous and crystalline phases in the matrix with the addition of 100% tuff. Zeolite phases grow at the expense of the amorphous gel-like matrix.
In general, the SEM photomicrographs indicate better geopolymerization and crystallization with the increase of tuff % as a result of its reactivity towards NaOH.

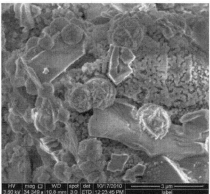

Figure 15. SEM photomicrograph of zeolites growth from gel (100% tuff).

Figure 16 illustrates the effect of high temperatures up to 1200°C on the compressive strength of the geopolymers' specimens from series (3) with 50 % tuff. The effect of the increase of temperature on the compressive strength has indicated the lowest values at 900 C (14 MPa) and the highest values (68.3 MPa) at 1100 C. At 1200 C the geopolymer specimen has almost completely melted.

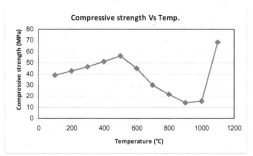

Figure 16. The effect of high temperatures up to 1200°C on the compressive strength of the geopolymer (series with 50% tuff).

The figure illustrates that the compressive strength increases up to 56 MPa with the increase of temperature to 500°C under dry conditions. Afterwards, the compressive strength decreases with the increase of the temperature up to 900°C (14.2 MPa). The mechanical performance remarkably increases again above 900 °C and reaches its maximum (68.3 MPa) at 1100°C before the geopolymer starts flowing out at 1200°C.

The XRD traces of geopolymers cured at 80 °C showed that the mineral phases are mainly kaolinite, muscovite, quartz and Na-Al silicate phases (Na- phillipsite and natrolite structures) [4; 10; current work].Figure 17 shows the effect of increase of temperature on mineral phases. Kaolinite major peaks (7 Å and 3.5 Å) disappear at 600°C as a result of dehydroxylation and change into metakaolinite.

The drop in the compressive strength values at a temperature higher than 600°C is related to the dehydroxylation of kaolinite and the disappearance of unstable Na-Al – silicate phases possibly as a result of melting and/or recrystallization (Figure 19). At 700°C the dehydroxylation of kaolinite is complete forming a poorly crystalline metakaolinite. Products produced at this temperature can act as a pozzolanic material.

At 1000°C amorphous phases are formed which undergo a transformation, with mullite recrystallizing in an amorphous glass at temperatures above 1100°C.

The increase in the compressive strength values above 800°C is related to the devitrification of amorphous fractions, healing the cracks formed due to the thermal treatment. Afterwards, these glassy phases can recrystallize (see appearance of sodalite and mullite). Sodalite is the result of reaction between metakaolinite and NaOH at high temperature. Heating the geopolymer at 1000°C for 24 h did not show a collapse of the texture and this is an empirical indication of refractoriness. The presence of sodalite and mullite refractory phases should at 1000°C made these geopolymers sufficiently refractory for continuous use up to this temperature. Figure 18 shows a heated specimen at 200°C with zeolites polymerized in the porous matrix. Residual collapsed amorphous kaolinite flakes and zeolite-rich matrix are present up to 900°C. At 900°C the material is partly vitreous, much less porous and the Na is held more strongly in the glassy and crystalline (nepheline - sodalite) phases (Figure 19). This explains the mechanical behavior. At 1000°C, 1100°C and 1200°C zeolitic and vitreous less porous textures are observed (Figure 20 – Figure 22).

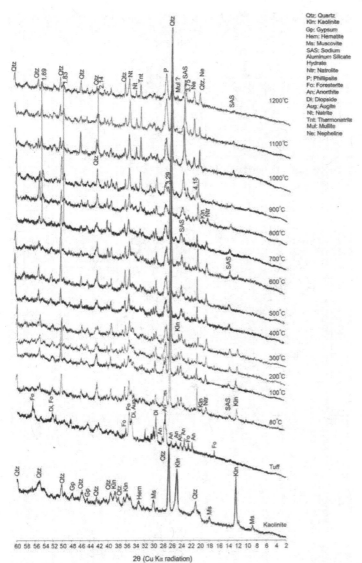

Figure 17. X-ray diffractograms illustrating the effect of temperature on mineral phases (series with 50% tuff).

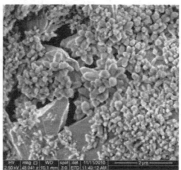

Figure 18. SEM photomicrograph of heated specimen at 200°C (50% tuff) illustrating zeolite growth from amorphous matrix.

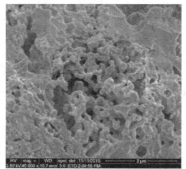

Figure 19. SEM photomicrograph of heated specimen at 900°C (50% tuff) illustrating porous crystalline and vitreous texture.

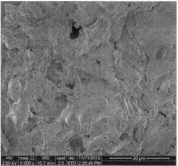

Figure 20. SEM photomicrograph of heated specimen at 1000°C (50% tuff) illustrating porous crystalline and vitreous texture.

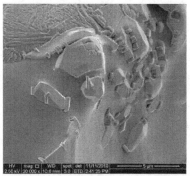

Figure 21. SEM photomicrograph of heated specimen at 1100°C (50% tuff) illustrating porous crystalline (zeolitic) and vitreous texture.

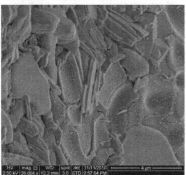

Figure 22. SEM photomicrograph of heated specimen at 1200°C (50% tuff) illustrating compact crystalline and vitreous texture.

CONCLUSIONS

Natural kaolinite with tuff and quartz fillers satisfies the criteria to be used as a precursor for the production of inexpensive and durable construction materials.

The specimens with 25% to 75 weight% pozzolanic tuff, replacing sand as filler, have maximum reaction with NaOH and have increased the compressive strength after 14 days to 47 MPa for dried samples and 21 MPa for immersed samples. The alkaline solution reacts with tuff to produce crystalline phases of Na-Al silicates and Na-Ca-Al silicates and low density products. The physical properties, SEM and XRD results have confirmed that the reaction with kaolinite, tuff and quartz cured at 80 °C for 14 hours was not complete. Zeolite phases grow at the expense of the amorphous gel-like matrix. The lower densities of the geopolymers obtained by the addition of 50 % tuff enable their use as an insulating material. Therefore, they can be used as insulating material and for applications where fire safety is required.

The effect of the increase of temperature on the compressive strength has indicated the lowest values (14.2 Mp) at 900 C and the highest values (68.3 Mpa) at 1100 C. The remarkable increase in compressive strength at 1100 C is related to the devitrification process of the molten part.

The SEM and XRD results have indicated the effect of increase of temperature on mineral phases, compressive strength and texture. The geopolymers' specimens exhibit maximum mechanical strength at 500 °C before it has dropped down to 14.2 MPa at 900 °C. The mechanical performance remarkably increased at a temperature higher than 900 °C and reached its maximum (68 MPa) at 1100 before the geopolymer has melted and recrystallized at 1200°C.

The presence of sodalite and mullite refractory phases should at 1100°C made these geopolymers sufficiently refractory for continuous use up to this temperature.

ACKNOWLEDGEMENT

The financial support of the Deanship of Scientific Research, University of Jordan of the project "Chemical stabilization of natural geomaterials for construction and industrial applications" is highly appreciated. The support of the Flemish (Belgium) Vlaamse Interuniversitaire Raad (VLIR, Contract ZEIN2006PR333) within the "Own Initiatives" program is gratefully acknowledged.

*Corresponding Author: khouryhn@ju.edu.jo

REFERENCES

[1] P.Duxson, A. Fernández-Jiménez, J.L Provis., G.C. Lukey, A. Palomo and J.S.J. van Deventer. Geopolymer technology: the current state of the art. Journal of Materials Science, 42, 2917-2933, (2007).

[2] J. Davidovits,"Chemistry of Geopolymeric Systems, Terminology," Geopolymere '99, Geopolymer International Conference, Proceedings, 30 June – 2 July, 1999, pp. 9-39, Saint-Quentin, France. Edited by J. Davidovits, R. Davidovits and C. James, Institute Geopolymere, Saint Quentin, France, (1999).

[3] H. Rahier, B. Van Mele, and J. Wastiels, Low Temperature Synthesized Aluminosilicate Glasses. Part B: Rheological Transformations during Low-Temperature Cure and High- Temperature Properties of a Model Compound, J. Mater. Sc., 31, 80-85, (1996).

[4] F. Slaty "Durability of Geopolymers Product from Jordanian Hiswa Clay" .PhD dissertation, University of Jordan, (2010).

[5](a) O.G. Ingles, Mechanism of clay stabilization with inorganic acids and alkalis, J. Soil Res. Aust. (1970) 81–95;

   (b) O.G. Ingles, Advances in soil stabilization, Rev. Pure Appl. Chem., Aust. (1968) 1961–1967.

[6] G. A. Patfoort, J. Wastiels, Use of Local Resources in Construction Materials, Vrije Universiteit Brussel, Brussels, Belgium, 1989, pp. 12–14.

[7] M. Rowles, B. O'Connor, Chemical optimization of the compressive strength of aluminosilicate geopolymers synthesized by sodium silicate activation of metakaolinite, J. Mater. Chem. 13 1161–1165, (2003).

[8] J. Davidovits, "Geopolymers: Man-Made Rock Geosynthesis and the Resulting Development of very Early High Strength Cement," Journal Materials Education, 16 [12] 91-139, (1994).

[9] ASTM D4318 D 4318, Standard Test Methods for Liquid Limit, Plastic Limit, and Plasticity Index of Soils, vol. 04.08, American Society for Testing and Materials, (2003)

[10] H. Rahier, F. Slaty, I. Al dabsheh, M. Al Shaaer, H. Khoury, M. Esaifan and J. Wastiels. Use of local raw materials for construction purposes (2010). In Science and Technology Vol. 69 pp 152-155. Trans Tech Publications, Switzerland.

[11] G.M. Gemerts, R. Mishre and J. Wastiels, Stabilization of Kaolinitic Soils from Suriname for Construction Purposes, Vrije Universiteit Brussel, Brussels, Belgium, (1989).

[12] J. Davidovits, Geopolymers and geopolymeric new material, J. Therm. Anal. 35, 429-441(1998).

[13] P. Duxson, G. C. Lukey, J. S. J. van Deventer, Physical evolution of Na-geopolymer derived from metakaolin up to 1000 °C. J Mater Sci 42:3044–3054 (2007).

[14] H. Khoury, M. AlShaaer (2009): Production of Building Products through Geopolymerization. GCREEDER. Proceedings Global Conference on Renewable and Energy Efficiency for Desert Regions. P 1-5, (2009).

[15] ASTM D 4318, Standard Test Methods for Liquid Limit, Plastic Limit, and Plasticity Index of Soils, vol. 04.08, American Society for Testing and Materials, (2003).

[16] R. I. Yousef, B. El-Eswed , M. Alshaaer, K. Fawwaz, H. Khoury, The influence of using Jordanian zeolitic tuff on the adsorption, physical, and mechanical properties of geopolymers products, J. Hazardous Materials, 165, Issues 1-3, 379-387, (2009)

[17] H. Khoury, Clays and Clay Minerals in Jordan, The University of Jordan, p.116, (2002).

[18] H. Khoury, Industrial Rocks and Minerals of Jordan. The University of Jordan, p. 289. (2006).

BOTTOM ASH-BASED GEOPOLYMER MATERIALS: MECHANICAL AND
ENVIRONMENTAL PROPERTIES

R. Onori*, J. Will**, A. Hoppe**, A. Polettini*, R. Pomi*, A.R. Boccaccini**
* Department of Civil and Environmental Engineering, University of Rome "La Sapienza"
Rome, Italy
**Department of Materials Science and Engineering, University of Erlangen-Nuremberg,
Erlangen, Germany

ABSTRACT
       The feasibility of geopolymer synthesis using incinerator bottom ash as the source of Si- and
Al- containing phases was tested in the present study. Alkaline activation at moderate temperatures
was used to force the rearrangement of the alumino-silicate matrix resulting in the production of a
geopolymeric material. Different mixtures were investigated for geopolymer production, which
differed in the Si/Al ratio (obtained upon addition of sodium silicate and metakaolin at varying
proportions) and the NaOH concentration in the alkaline medium. The obtained geopolymeric
materials were characterized in terms of physical and mechanical properties, mineralogical and
microstructural characteristics as well as chemical durability. Wide variations in the main
characteristics of the obtained products were observed, indicating a significant influence of the process
variables on the evolution of the geopolymerization reactions. While the physical and mechanical
properties correlated well with the Si/Al ratio, no clear trend was found with the NaOH concentration;
however, although the strength development upon geopolymerization was clear, the experimental data
suggested that the optimal Si/Al ratio was probably beyond the investigated range. SEM, FT-IR, TG
and leaching test data showed that, although the onset of the geopolymer-formation reactions was
clear, the degree of matrix restructuring was relatively poor, indicating the need for further
investigation to promote the development of the geopolymerization reactions.

INTRODUCTION
       Bottom ashes (BAs) from waste incineration are generated in relatively large amounts (> 90%
of the total solid residues mass) and account for 15-25% by mass of the original waste. BAs are also
the residues with technical properties most suited for utilization. The similarities in physical
characteristics and major composition of granular construction materials and BA make this residue
virtually suited for recycling as a substitute for natural aggregate; construction and building materials
are the primary utilization route for BAs in Europe[1]. All the mentioned utilization options rely on the
implicit assumption of the predominantly inert characteristics of BA, however, the chemical reactivity
of BA under natural conditions has also been widely assessed. Thus, advantage may be taken from the
presence of reactive compounds in BA to improve the materials' properties for utilization. Different
processing methods have been investigated to improve the mechanical properties of BA in view of
engineering applications, which are mostly based on preliminary activation of the material by means of
mechanical, chemical or thermal methods (or a combination of the three).
       In our previous studies[2-6] we applied a number of processes of different nature to transform
incinerator BA in a valuable material for various applications. Namely, the studies focused on the
production of: 1) sintered ceramics to be used for manufacturing of tiles and linings for pipes and ducts
or as an aggregate material; 2) zeolitized materials for the treatment of high-strength industrial
wastewaters; 3) pozzolanic materials for the formulation of blended cements. In the present work, we
intend to gain insights into the feasibility of production of geopolymeric materials from incinerator
BA, considering that the material is known to be composed of amorphous and vitreous phases, the
most abundant constituents being Ca-, Si- and Al-containing minerals. While other kinds of inorganic
industrial wastes (coal/lignite fly ash[7-15], biomass ash[16], coal bottom ash[9,17], blast furnace slag[18,19], air

71

pollution control ash from waste incineration[20] and wastewater sludge[21]) have been investigated in previous studies for similar applications, to the authors' knowledge no specific study on geopolymer production from incinerator BA has been conducted so far.

MATERIALS AND METHODS

Waste incinerator BA from an Italian grate-type RDF incinerator was used as the starting material for geopolymer synthesis. Fresh BA was sampled at the quenching unit outlet, homogenized by quartering and characterized for elemental composition and anion content. The elemental composition was determined using a Perkin-Elmer 3030B atomic absorption spectrometer after triplicate alkaline digestions at 1050 °C in platinum crucibles using lithium tetraborate as the melting agent. Anions were analyzed using a Metrohm 761 Compact IC ion chromatographer after dissolution according to the Italian UNI 8520 methods. Microstructure was investigated with a Philips XL-30 SEM analyzer, operating at 25 kV with a spot size of 200 nm, a tilt angle of 35°, a take-off angle of 61.34° and equipped with an EDAX DX-4 energy-dispersive spectrometer operated at a count rate of 1200 cps and a live time of 50 s. Samples were first impregnated with an ultra-low viscosity resin, then polished and carbon-coated under vacuum.

BA pre-treatment involved ball-milling the dried as-received material to a final mean particle size < 425 μm. The alkaline activator used for geopolymerization included a sodium silicate solution (SiO_2 = 26.91%, Na_2O = 8.68%; SiO_2/Na_2O = 3.1 w/w) and an NaOH solution (10–15.5 M), which were mixed in different proportions to provide the desired reacting medium. Metakaolin (MK) was used to adjust the Si/Al ratio in the desired range of values; metakaolin was obtained from commercial-grade kaolin ($Al_2O_3 \cdot 2SiO_2 \cdot 2H_2O$; CAS no. 1332-58-7; > 95% kaolinite content) by calcination at 800°C for 2 hours in air. Milled BA was mixed with the alkaline activator at a solid-to-liquid (S/L) ratio of 4 kg/l except for pastes with the highest metakaolin content, where additional water was required to ensure a suitable workability, resulting in a S/L ratio of 3.3.

Different pastes were tested for the geopolymerization process by varying the Si/Al molar ratio (1.28–2.29) and the NaOH concentration (5–9 M) in the alkaline medium, as reported in Table I (sample codes include the values of the two parameters in each mixture). The paste preparation procedure involved hand-mixing of the dry BA and metakaolin for 2 minutes, followed by hand-mixing for additional 7 minutes of the resulting dry mixture with previously blended Na silicate and NaOH solutions. The resulting pastes were cast in 20-mm diameter, 45-mm height cylindrical steel moulds, manually compacted with a piston and de-moulded afterwards. The cylindrical specimens were transferred to an oven where they were heated at 75 °C for 24 hours; such operating values were selected on the basis of the results from preliminary tests. After the heat treatment the specimens were wrapped in parafilm foils and cured at room temperature for 7 days.

Table I. Composition of pastes used for the geopolymerization experiments

Sample code	Mixture components					S/L ratio	Final values		
	BA	MK	Na silicate sol.	NaOH sol.			Si/Al	NaOH	Na/Al
	(g)	(g)	(ml)	(M)	(ml)	(g/ml)	(mol/mol)	(M)	(mol/mol)
S-2.29-5	16	4	2.5	10	2.5	4	2.29	5	0.50
S-1.69-5	12	8	2.5	10	2.5	4	1.69	5	0.33
S-1.28-5	5.2	14.8	3	10	3	3.3	1.28	5	0.25
S-2.29-6	15.8	4.2	3	15	2	4	2.29	6	0.57
S-1.69-6	11.7	8.3	3	15	2	4	1.69	6	0.38
S-1.28-6	4.6	15.4	3.5	14.5	2.5	3.3	1.28	6	0.29
S-2.29-7.5	16	4	2.5	15	2.5	4	2.29	7.5	0.69
S-1.69-7.5	12	8	2.5	15	2.5	4	1.69	7.5	0.46
S-1.28-7.5	5.2	14.8	3	15	3	3.3	1.28	7.5	0.35
S-2.29-9	16.2	3.8	2	15	3	4	2.29	9	0.82
S-1.69-9	12.3	7.7	2	15	3	4	1.69	9	0.55
S-1.28-9	5.7	14.3	2.5	15.5	3.5	3.3	1.28	9	0.42

After 7 days, the products were tested for physical, mechanical, mineralogical, microstructural and leaching properties. Such a curing time was selected on the basis of preliminary experiments, which evidenced a gain in mechanical strength from 1 to 7 days also accompanied by a reduction in variance of measured data, while a remarkably lower increase on a longer term. Of course, however, the main gain in strength was caused in the short term by the alkaline treatment applied. Physical and mechanical characterization involved triplicate measurements of bulk density (volume displacement principle), total porosity (pycnometric measurement) and unconfined compressive strength (UCS). Mineralogy and microstructure were investigated through SEM/EDAX, FT-IR and TGA/DTA analyses; while SEM observations were carried out on specimen fragments obtained from mechanical testing, the other analytical techniques were applied on powdered samples. SEM analyses were conducted as already described above. FT-IR absorbance spectra were collected with an Impact 420 Nicolet instrument in the wavenumber range 400–4000 cm^{-1} with a resolution of 4 cm^{-1} on pellets with KBr. Simultaneous TGA/DTA analyses were performed on a Netzsch STA 409 C/CD instrument at a heating rate of 10 °C/min in static air over a temperature range of 20–1000 °C. The EN 14429 (acid neutralization capacity, ANC) leaching test was applied after mechanical strength testing on ball-milled (<425 μm) specimens to assess the chemical durability of the materials.

RESULTS AND DISCUSSION

Chemical, microstructural and mechanical characterisation

The elemental composition of the BA used is reported in Table II. The major constituents of the untreated bottom ash are: Ca (27.5% dry wt.), Si (15.0% dry wt.), soluble chloride (5.4% dry wt.) and Al (4.1% dry wt.). Appreciable amounts of trace metals were also detected in the material, the highest contents being displayed by Cu (5900 mg/kg), Pb (1400 mg/kg) and Zn (6200 mg/kg).

Table II. Concentrations (average values ± standard error) of major elements, trace metals and anionic species in the untreated BA

Element/Anion	Concentration (mg/kg)	Element/Anion	Concentration (mg/kg)
Al	41200 ± 2%	Mo	30 ± 4%
As	3.4 ± 6%	Na	17500 ± 1%
Ca	270000 ± 7%	Ni	170 ± 22%
Cd	8 ± 15%	Pb	1400 ± 12%
Cr	390 ± 4%	Sb	140 ± 15%
Cu	5900 ± 16%	Si	150000 ± 3%
Fe	29400 ± 3%	Zn	6200 ± 6%
K	3870 ± 2%	Cl$^-$	54000 ± 5%
Mg	18900 ± 4%	SO$_4^{2-}$	4100 ± 5%
Mn	710 ± 5%		

The grain size distribution of the milled BA was as follows: 250 μm ≤ Φ < 425 μm = 37%, 150 μm ≤ Φ < 250 μm = 21%, 100 μm ≤ Φ < 150 μm = 15%, 75 μm ≤ Φ < 100 μm = 17%, 63 μm ≤ Φ < 75 μm = 7.5%, 38 μm ≤ Φ < 63 μm = 1.8%, Φ < 38 μm = 0.7%. The SEM micrograph (100 magnification) of the untreated milled BA is shown in Figure 1, which indicates a microstructure with loose particles with large areas characterized by a dark gray level intermixed with lighter gray zones of more limited extension. Both areas were analyzed for elemental composition through EDAX analyses. Dark gray areas mainly contained Ca (16 mol%), Si (13.5 mol%) and Al (5.4 mol%), which could roughly be described by the presence of CaO, SiO$_2$ and Al$_2$O$_3$ at molar ratios of 6:5:1; lower contents of C (4.6 mol%), Na (3.7 mol%), Cl (1.9 mol%) and Mg (1.3 mol%) were also detected. Light gray zones mainly contained Si (24.4 mol%), Na (9.8 mol%) and Ca (7.3 mol%), with smaller concentrations of C (3.9 mol%), Al (2.8 mol%) and Mg (1.4 mol%).

The results from mechanical characterization are reported in Figure 2 in terms of UCS values and associated standard deviation. Depending on paste composition, large variations in the mechanical properties were observed, with values ranging from a minimum of < 0.1 MPa for sample S-1.28-7.5 to a maximum of 7.4 MPa for sample S-2.29-7.5. The influence of the Si/Al and Na/Al ratios on UCS is depicted in Figures 3 a) and b). An increase in both the Si/Al and Na/Al ratios appeared to improve compressive strength, although at fixed values of the Na/Al ratio an increase in the NaOH concentration in the alkaline reaction medium seems to exert a negative effect on mechanical strength. For the Si/Al ratio, the results shown in Figure 3 a) appear to indicate that the optimum for mechanical strength development should lie beyond the investigated range. While the measured UCS values were relatively weak at the operating conditions tested, the trend reported in Figure 3 a) suggests that improved mechanical characteristics may be obtained when increasing the Si/Al molar ratio above 2.29. Previous studies of geopolymerization of waste materials[7-21] have shown that a very wide range of mechanical strength values can be obtained depending on the initial material composition and processing conditions.

As for the relative contribution of BA and MK to strength development, although with the available experimental data it was not possible to derive any specific conclusion, it appears that UCS benefited from an increase in the bottom ash content in the mixtures (data not reported graphically here), indicating the reactivity of the material in the alkaline medium. However, a minimum amount of MK was also found to be required for the investigated mixtures to attain suitable mechanical properties.

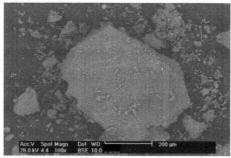

Figure 1. SEM micrographs of polished surfaces for untreated BA.

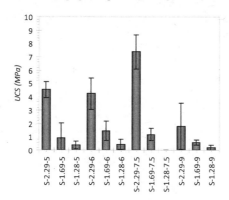

Figure 2. Results of UCS testing on geopolymerized materials (error bars are drawn at one standard deviation of values).

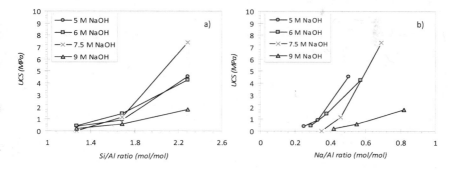

Figure 3. Relationship between UCS and (a) Si/Al and (b) Na/Al ratio.

For a more detailed physical and mineralogical/microstructural characterization and to gain knowledge on the microstructure-mechanical strength relationship, four samples were selected among those displaying the best and worst mechanical behaviour; these included S-2.29-7.5 (UCS = 7.38 ± 1.29 MPa), S-2.29-5 (UCS = 4.52 ± 0.59 MPa), S-1.69-7.5 (UCS = 1.18 ± 0.45 MPa) and S-1.28-7.5 (UCS < 0.1 MPa), which differed in the Si/Al ratio and the NaOH concentration. The unit weight and total porosity values for such samples are reported in Table III. Unit weight was found to vary from 10.5 to 15.6 kN/m^3, while total porosity ranged from 40.2 to 66.5%, and both parameters displayed a good correlation with mechanical strength data.

Table III. Results of physical characterization for selected samples (mean values ± standard deviation)

Sample code	UCS (MPa)	Unit weight (kN/m^3)	Porosity (%)
S-2.29-7.5	7.38 ± 1.29	14.96 ± 0.12	42.8 ± 1.2
S-2.29-5	4.52 ± 0.59	15.57 ± 0.15	40.2 ± 0.5
S-1.69-7.5	1.18 ± 0.45	12.02 ± 0.21	58.9 ± 0.9
S-1.28-7.5	< 0.1	10.54 ± 0.24	66.5 ± 1.2

The SEM micrographs (200 magnification) of polished fragments for samples S-2.29-7.5 and S-2.29-5 (Figure 4) suggest that individual unreacted particles of different size were still present in the material and were surrounded by a layer of small sheets of reacted solids. This effect has been also observed in geopolymers produced using plasma vitrified incinerator air pollution control residues[20]. In the present case, it is apparent that the external layer of reacting material filled the original pores and acted by forming bridges and interconnections between the original grains (Figure 4). The fact that geopolymerization reactions had occurred within the material appears to be indicated by the relatively low pore volume and small void size visible in SEM pictures. Visual observation of the amount and size of pores in the investigated samples also appears to confirm the results from total porosity measurements which indicated that sample S-2.29-7.5 was more porous than sample S-2.29-5. EDAX spot analyses conducted on the 10-μm square areas indicated in Figure 4 showed the main presence of Si (20.9 and 18.0 mol% for samples S-2.29-7.5 and S-2.29-5, resp.) and Al (19.8 and 15.7 mol% for samples S-2.29-7.5 and S-2.29-5, resp.), corresponding to Si/Al molar ratios of 1.05 and 1.15. The Si/Al ratio for these samples thus decreased from 2.29 mol/mol to a final value of ~1.0 mol/mol, which may be explained considering that polysialate-type geopolymers, having an Si/Al ratio of 1[22], were formed under the treatment conditions applied.

Sample S-2.29-5 was also found to have a significant Na content (10.8 mol%), which was also appreciably higher than that of sample S-2.29-7.5.

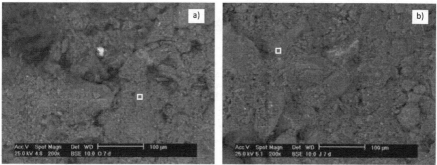

Figure 4. SEM micrographs of fracture surfaces for samples a) S-2.29-7.5 and b) S-2.29-5.

The results of infrared spectroscopy characterization are reported in Figure 5 a) for the untreated BA and the four BA-based geopolymeric materials mentioned above. Six main adsorption bands can be identified in the FT-IR spectra of treated materials, namely: a band at 450–470 cm^{-1} associated to in-plane bending vibration of Si–O–Si[23,26] (referred to as peak 1), a second band of low intensity at 710–730 cm^{-1} related to symmetric stretching vibration of Si–O–Si or Si–O–Al bonds, a main band at 980–1040 cm^{-1} associated to asymmetric stretching vibrations of Si–O/Al–O bonds[23-26] (peak 2), a band at ~1450 cm^{-1} due to the presence of O–C–O stretching vibration in carbonate groups[7,23,24,27] (peak 3), two broad bands at 1640–1660 cm^{-1} and 3440–3460 cm^{-1} related to stretching and deformation vibrations of OH and H–O–H groups[9,23,27,28] (peaks 4 and 5). For the untreated BA, a sharp band (although of low intensity) at ~3640 cm^{-1} was also identified (peak 6) and associated to O–H stretching in the Ca(OH)$_2$ structure[29].

According to previous literature studies[23-26], the peaks centered at 980–1040 cm^{-1} and 450–470 cm^{-1} were assumed as the main molecular vibration fingerprints of geopolymeric materials. While the latter was not present in the original BA and was visible in the treated products, the former appeared in the treated materials with a different shape and intensity if compared to the untreated ash. The absorption characteristics in the range of peak 2 suggest that FT-IR spectra display broad and asymmetric curves, which likely result from the overlapping of several characteristic vibrational bands[30]; it is thus possible that not only stretching vibrations of Si–O/Al–O bonds were associated to the observed IR peaks, but that additional vibration modes were also hidden in the spectra. However, a quantitative derivation of such additional characteristic bands has not been performed in the present study. Considering the main peak identified as peak 2, its position tended to shift from wavenumbers of 1020–1040 cm^{-1} (samples S-1.69-7.5 and S-1.28-7.5) towards lower values in the range 985–1007 cm^{-1} (samples S-2.29-5 and S-2.29-7.5), which may be taken as an indication of the increase in the degree of geopolymerization associated to the inclusion of tetrahedrally-coordinated Al in the Si–O–Si skeletal structure[7,12,24,30,31]. The presence of Al in the silicate network may also be confirmed by the appearance of a shoulder in the FT-IR spectra at ~1200 cm^{-1}, which has been related to aluminosilicate structures[32]. On the other hand, according to Lecomte et al.[33], the occurrence of a band at around 710–730 cm^{-1} may provide an indication of the fact that a relatively poorly polymerized structure was formed in the investigated materials, since this band has been associated to silicooxygen rings with a low number of structural members. It may thus be hypothesized that in the investigated materials an aluminosilicate network possibly coexisted with C–S–H or C–A–S–H type phases, as indicated by numerous studies on the effects of Ca on the structure of geopolymeric materials[34-36]. These phases may have been formed from the reaction of Ca originating from BA with dissolved Si and Al; the disappearance of some reactive Ca from the original material may be confirmed by the fact that the peak at ~3640 cm^{-1} in the FT-IR curve associated to Ca(OH)$_2$ could not be detected in the treated material, indicating possible reactions of Ca to form (alumino)silicate phases.

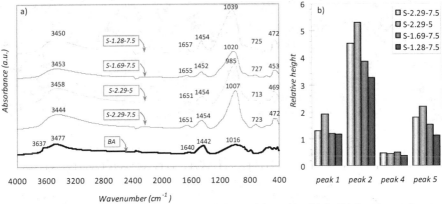

Figure 5. FT-IR spectra (a) and comparison of relative peak intensities (b) for BA-based geopolymers.

A comparison between the intensities of FT-IR peaks for the four BA-based geopolymers is provided in Figure 5 b). Data for each sample were calculated as relative intensities of the peaks of interest normalised to the intensity of peak 3, assuming that the degree of carbonation was the same for all samples tested, so that the carbonate peak should have the same height. The most evident difference in peak height could be observed for peak 2, with samples S-2.29-5 and S-2.29-7.5 displaying higher peak intensities than the other two samples, in agreement with the higher strengths measured; however, the fact that sample S-2.29-7.5 showed a higher UCS than the former could not be related to the information provided by FT-IR data. In general, FT-IR data showed that sample S-2.29-5 displayed higher peak intensities than the other samples. The fact that the height of peak 5, associated to absorbed water, followed the same trend already noted for peak 2 may be considered as an indication of the higher degree of hydration attained for samples S-2.29-5 and S-2.29-7.5, as opposed to the other materials investigated.

Another interesting feature noted in the FT-IR spectra was the fact that a small peak appearing at ~800 cm^{-1} for the raw BA, which was attributed to Si–O–Al vibration[23], disappeared in the treated materials and was replaced by several weaker bands at lower wavenumbers, ranging from 700 to 800 cm^{-1}. This may be interpreted assuming that decomposition of Al-containing chemical structures occurred during geopolymerization and that the resulting Al was afterwards incorporated in alumino-silicate structures indicated by peak 2.

The thermal analysis results are reported in Figure 6. The thermograms displayed basically four regions characterized by different amounts and rates of mass loss in the following ranges: 20–240 °C, 240–380 °C, 380–700 °C and >700 °C. Within the first region, a major weight loss occurred, accounting for 58–75% of the total weight loss recorded over the temperature range investigated. As outlined by numerous studies[7,12,37-40], a large endothermic peak was observed at temperatures of 120–130 °C, which was related to sample dehydration, with loss of absorbed and loosely bound water. As reported in Table IV, the weight loss associated to the first temperature region ranged from 4.4 to 5.3% and it was larger for samples S-1.69-7.5 and S-1.28-7.5, for which a poorer geopolymerization degree was also indicated by the other characterization analyses.

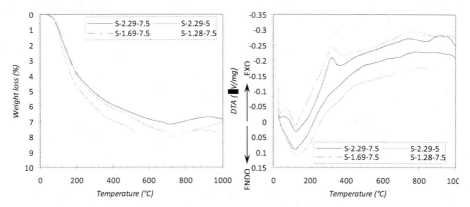

Figure 6. TGA and DTA results for BA-based geopolymers.

Further weight decrease was recorded, although at a lower rate (as also noted by other investigators[38,40]), in the second temperature region and was associated to sample dehydroxylation[7,37]; the mass loss values ranged from 1.4 to 1.8%. However, in the DTA patterns of samples S-1.69-7.5 and S-1.28-7.5, an exothermic peak was identified at about 330 °C, with an associated weight loss of 1.6 and 1.4%, respectively, which indicates the presence of a different phase for such samples if compared to the other materials investigated. Based on the information available, however, no tentative association with any specific phase, the occurrence of which could reasonably be expected only in samples S-1.69-7.5 and S-1.28-7.5, could be made. If the weight loss associated to the decomposition of this unknown phase is subtracted from the overall weight decrease in the temperature region 240–380 °C, net mass loss values are obtained, as shown in Table IV, which suggested a higher degree of hydration for samples S-2.29-7.5 and S-2.29-5, possibly confirming the larger geopolymerization degree attained for such materials.

An additional, much slower, mass loss was observed between 380 and 700 °C, which mirrors a more stable thermal behaviour of the materials at temperatures above 380 °C and confirms the results obtained from other studies[38,41]. Finally, a small weight gain (~1%) was observed in the fourth temperature region (> 700 °C), which may be due to oxidative transformations, although the identification of the species involved has not been possible.

Table IV. Weight loss for different temperature regions as derived from the TGA curves in Figure 6

| | Sample code | | | |
Temperature range	S-2.29-7.5	S-2.29-5	S-1.69-7.5	S-1.28-7.5
0-240 °C	4.42	4.57	4.81	5.27
240-380 °C	1.38	1.45	1.83	1.57
240-380 °C (net)	1.38	1.45	0.21	0.19
380-700 °C	1.36	1.74	1.50	1.24
700-1000 °C	-0.33	0.05	-0.56	-1.03

Chemical durability

The ANC curves of the untreated BA and the BA-based geopolymers are shown in Figure 7.

Significant changes in the buffering capacity were observed in comparison to the raw ash and between the geopolymerized samples. Due to the high alkali amount added to the mixtures, the natural pH of the original BA (11.84) was found to increase by almost two units in the BA-based geopolymers, reaching values in the range 13.02–13.33. The different mineralogical composition of the untreated and treated materials is mirrored by the different shapes of the titration curves: while the raw BA displayed a sort of plateau in the pH range 10–12, the curves of the treated materials were steeper in this pH range, indicating no significant buffering by mineral phases. The acid buffering capacity of the treated material was generally lower than that of the original BA within the whole range investigated: for the purpose of comparison, while the ANC to pH 8 was 3.2 meq H^+/g for the raw BA, it was reduced to 1.6 meq H^+/g (sample S-1.28-7.5) – 2.9 meq H^+/g (sample S-2.29-7.5). Furthermore, samples S-2.29-5 and S-2.29-7.5 displayed similar ANC curves in the entire pH range investigated and showed stronger buffering capacities with respect to the other two samples; these also appreciably differed from each other, with sample S-1.28-7.5 displaying the steepest curve. The lower buffering capacity of samples S-1.28-7.5 and S-1.69-7.5 is assumed to confirm the lower degree of geopolymerization reactions occurred for these materials as indicated by the previous analyses.

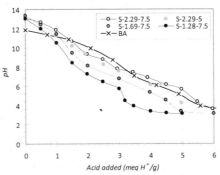

Figure 7. ANC curves for BA-based geopolymers as compared to the untreated BA.

CONCLUSIONS

In the present work, a number of mixtures obtained from BA with the addition of metakaolin, sodium silicate and sodium hydroxide were subjected to a low-temperature thermal treatment in order to induce the formation of geopolymeric materials. Physical, mechanical, microstructural and mineralogical analyses were applied in order to derive information about the influence of the Si/Al ratio and the amount of added NaOH on the main properties of the obtained products. Depending on the process conditions adopted, wide variations in the main characteristics of the materials were observed, indicating a significant influence of the selected process variables on the evolution of the geopolymerization reactions. While the physical and mechanical properties were found to be well correlated with the Si/Al ratio, no clear trend was found with the NaOH concentration in the alkaline medium. Mechanical strength data also suggested that further investigation on the Si and Al content is required to identify the optimal range of values for inducing the highest possible mechanical properties. SEM, FT-IR, TG and chemical durability (ANC) data showed that, although the onset of the geopolymer-formation reactions was clearly confirmed, the degree of matrix restructuring was relatively poor, in agreement with the suggestions provided by the analysis of the mechanical properties.

REFERENCES

[1] T. Astrup, P. Lechner, A. Polettini, R. Pomi, T. Van Gerven, A. van Zomeren, State-of-the-art and outlook on management of waste-to-energy bottom ashes. Part 2: Utilization. Proc. Sardinia 2007, Eleventh International Waste Management and Landfill Symposium, 2007.

[2] C.R. Cheeseman, S. Monteiro da Rocha, C. Sollars, S. Bethanis, and A.R. Boccaccini, Ceramic processing of incinerator bottom ash, *Waste Manage.*, **23** 907–916 (2003).

[3] A. Polettini, R. Pomi, S. Lo Mastro, and S. Sampieri, Hydrothermal zeolitization of MSWI bottom ash, Proc. WASCON 2006, 6th International Conference on the Environmental and Technical Implications of Construction with Alternative Materials. Science and Engineering of Recycling for Environmental Protection, 439–50.

[4] A. Polettini, R. Pomi, and G. Carcani, The effect of Na and Ca salts on MSWI bottom ash activation for reuse as a pozzolanic admixture, *Resour. Conserv. Recy.*, **43**, 404–18 (2005).

[5] A. Polettini, R. Pomi, and E. Fortuna, Chemical activation in view of MSWI bottom ash recycling in cement-based systems. *J. Hazard. Mater.*, **162**, 1292–9 (2009).

[6] R. Onori, A. Polettini, and R. Pomi, Mechanical properties and leaching modeling of activated incinerator bottom ash in Portland cement blends, *Waste Manage.*, **31**, 298–310 (2011).

[7] S. Andini, R. Cioffi, F. Colangelo, T. Grieco, F. Montagnaro, and L. Santoro, Coal fly ash as raw material for the manufacture of geopolymer-based products, *Waste Manage.*, **28**, 416-23 (2008).

[8] P. Chindaprasirt, T. Chareerat, and V. Sirivivatnanon, Workability and strength of coarse high calcium fly ash geopolymer, *Cement Concrete Comp.*, **29**, 224-9 (2006).

[9] P. Chindaprasirt, C. Jaturapitakkul, W. Chalee, and U. Rattanasak, Comparative study on the characteristics of fly ash and bottom ash geopolymers, *Waste Manage.*, 29, 539-43 (2009).

[10] S. Kumar, and R. Kumar, Mechanical activation of fly ash: effect on reaction, structure and properties of resulting geopolymer, *Ceram. Int.*, **37**, 533-41 (2011).

[11] D. Panias, I.P. Giannopoulou, and T. Perraki, Effect of synthesis parameters on the mechanical properties of fly ash-based geopolymers, *Colloid. Surface A*, **301**, 246-54 (2007).

[12] F. Škvára, L. Kopecký, V. Šmilauer, and Z. Bittnar, Material and structural characterization of alkali activated low-calcium brown coal fly ash, *J. Hazard. Mater.*, **168**, 711-20 (2009).

[13] J. Temuujin, and A. van Riessen, Effect of fly ash preliminary calcination on the properties of geopolymer, *J. Hazard. Mater.*, **164**, 634-9 (2009).

[14] J. Temuujin, A. van Riessen, and K.J.D. MacKenzie, Preparation and characterisation of fly ash based geopolymer mortars, *Constr. Build. Mater.*, **24**, 1906-10 (2010).

[15] J.G.S. van Jaarsveld, J.S.J. van Deventer, and G.C. Lukey, The effect of composition and temperature on the properties of fly ash- and kaolinite-based geopolymers, *Chem. Eng. J.*, **89**, 63-73 (2002).

[16] S. Songpiriyakij, T. Kubprasit, C. Jaturapitakkul, and P. Chindaprasirt, Compressive strength and degree of reaction of biomass- and fly ash-based geopolymer, *Constr. Build. Mater.*, **24**, 236-40 (2009).

[17] H. Xu, Q. Li, L. Shen, W. Wang, and J. Zhai, Synthesis of thermostable geopolymer from circulating fluidized bed combustion (CBFC) bottom ashes, *J. Hazard. Mater.*, **175**, 198-204 (2010).

[18] T.W. Cheng, and J. P. Chiu, Fire-resistant geopolymer produced by granulated blast furnace slag, *Miner. Eng.*, **16**, 205-210 (2003).

[19] J.E. Oh, P.J.M. Monteiro, S.S. Jun, S. Choi, and S.M. Clark, The evolution of strength and crystalline phases for alkali-activated ground blast furnace slag and fly ash-based geopolymers, *Cement Concrete Res.*, **40**, 189-96 (2010).

[20] I. Kourti, D.A. Rani, D.Deegan, A.R. Boccaccini, and C.R. Cheeseman, Production of geopolymers using glass produced from DC plasma treatment of air pollution control (APC) residues, *J. Hazard. Mater.*, **176**, 704–9 (2010).

[21] C. Lampris, R. Lupo, and C.R. Cheeseman, Geopolymerisation of silt generated from construction and demolition waste washing plants, *Waste Manage.*, **29**, 368–73 (2009).

[22] J. Davidovits, Geopolymer chemistry and applications, Institut Géopolymère, Saint-Quentin, France, 2nd ed., 2008.

[23] V.F.F. Barbosa, K.J.D MacKenzie, and C. Thaumaturgo, Synthesis and characterisation of materials based on inorganic polymers of alumina and silica: sodium polysialate polymers, *J. Inorg. Mater.*, **2**, 309-17 (2000).

[24] W. Mozgawa, and J. Deja, Spectroscopic studies of alkaline activated slag geopolymers, *J. Mol. Struct.*, **924-926**, 434-41 (2009).

[25] X. Guo, H. Shi, W. Dick, Compressive strength and microstructural characteristics of class C fly ash geopolymer, *Cement Concrete Comp.*, **32**, 142–7 (2010).

[26] J.W. Phair, and J.S.J. Van Deventer, Effect of the silica activator pH on the microstructural characteristics of waste-based geopolymers, *Int. J. Miner. Process.*, **66**, 121-43 (2002).

[27] W.K.W Lee, and J.S.J. Van Deventer, The effects of inorganic salt contamination on the strength and durability of geopolymers, *Colloid. Surface A*, **211**, 115-26 (2002).

[28] A. Palomo, M.T. Blanco-Varela, M.L. Granizo, F. Puertas, T. Vazquez, M.W. Grutzeck, Chemical stability of cementitious materials based on metakaolin, *Cement Concrete Res.*, **29**, 997–1004 (1999).

[29] M. Yousuf, A. Mollah, P. Palta, T.R. Hess, R.K. Vempati, and D.L. Cocke, Chemical and physical effects of sodium lignosulfonate superplasticizer on the hydration of portland cement and solidification/stabilization consequences, *Cement Concrete Res.*, **25**, 671–82 (1995).

[30] P. Rovnaník, Effect of curing temperature on the development of hard structure of metakaolin-based geopolymer, *Constr. Build. Mater.*, **24**, 1176-83 (2010).

[31] C.A Rees, J.L. Provis, G.C. Luckey, and J.S.J. Van Deventer, The mechanism of geopolymer gel formation investigated through seeded nucleation, *Colloid. Surface A*, **318**, 97–105 (2008).

[32] A. Hidalgo, S. Petit, C. Domingo, C. Alonso, and C. Andrade, Microstructural characterization of leaching effects in cement pastes due to neutralisation of their alkaline nature. Part I: Portland cement pastes, *Cement Concrete Res.*, **37**, 63–70 (2007).

[33] I. Lecomte, C. Henrist, M. Liégeois, F. Maseri, A. Rulmont, and R. Cloots, (Micro)-structural comparison between geopolymers, alkali-activated slag and Portland cement, *J. Eur. Ceram. Soc*, **26**, 3789–97 (2006).

[34] K. Dombrowski, A. Buchwald, and M. Wail, The influence of calcium content on the structure and thermal performance of fly ash based geopolymers, *J. Mater. Sci*, **42**, 3033-43 (2007).

[35] K.J.D. MacKenzie, M.E. Smith, and A. Wong, A multinuclear MAS NMR study of calcium-containing aluminosilicate inorganic polymers, *J. Mater. Chem*, **17**, 5090-6 (2007).

[36] J. Temuujin, A. van Riessen, and R. Williams, Influence of calcium compounds on the mechanical properties of fly ash geopolymer pastes, *J. Hazard. Mater.*, **167**, 82-8 (2009).

[37] P. Duxson, G.C. Lukey, J.S.J. van Deventer, Thermal evolution of metakaolin geopolymers: Part 1 – Physical evolution, *J. Non-Cryst. Solids*, **352**, 5541–55 (2006).

[38] D.L.Y. Kong, and J.G. Sanjayan, Effect of elevated temperatures on geopolymer paste, mortar and concrete, *Cement Concrete Res.*, **40**, 334–9 (2010).

[39] Z. Pan, and J.G. Sanjayan, Stress–strain behaviour and abrupt loss of stiffness of geopolymer at elevated temperatures, *Cement Concrete Comp.*, **32**, 657–64 (2010).

[40] W.D.A. Rickard, and A. van Riessen, Thermal character of geopolymers synthesized from class f fly ash containing high concentrations of iron and -quartz, *Int. J. Appl. Ceram. Technol.*, **7**, 81–8 (2010).

[41] E. Prud'homme, P. Michaud, E. Joussein, C. Peyratout, A. Smith, S. Arrii-Clacens, J.M. Clacens, and S. Rossignol, Silica fume as porogent agent in geo-materials at low temperature, *J. Eur. Ceram. Soc.*, **7**, 1641–8 (2010).

PRODUCTION OF GEOPOLYMERS FROM UNTREATED KAOLINITE

H. Rahier[1], M. Esaifan[2], I. Aldabsheh[3], F. Slatyi[3] , H. Khoury[3], J. Wastiels[2],

[1]Department of Physical Chemistry and Polymer Science, Vrije Universiteit Brussel, Pleinlaan 2, 1050 Brussel, Belgium

[2]Department of Mechanics of Materials and Constructions, Vrije Universiteit Brussel, Pleinlaan 2, 1050 Brussel, Belgium

[3]Materials Research Laboratory, University of Jordan, Amman, Jordan.

Corresponding author H. Rahier: Email hrahier@vub.ac.be

ABSTRACT

Three Jordanian raw materials, namely Jordanian Hiswa kaolinite, Jordanian volcanic tuff and Jordanian smectite rich clay, are investigated on their potential for use as an alkali activated cement. A dissolution study of these materials already indicates that the tuff sample reacts slower than the two clay materials. DSC shows that the kaolinite is the most reactive material followed by the smectite rich clay. The tuff is much less reactive. The kaolinite is chosen for producing geopolymer specimens. The optimal amount of NaOH to be added is found to be 16 parts by weight compared to the kaolinite. These specimens have a compressive strength of 33 MPa under dry conditions, after curing for 24h at 80°C. TGA can be used for checking how much kaolinite remains after the geopolymerization.

INTRODUCTION

Since more than 25 years, effort has been carried out at Vrije Universiteit Brussels to produce construction materials, starting from local raw materials. A technique often used to obtain concrete like materials is alkali activation[1-6]. With this technique for instance geopolymers can be produced[7,8]. Geopolymers have the benefit that they have a smaller environmental impact than does concrete[9], but also the main raw material, for instance kaolinite, is often locally available[10,11]. To further minimize the production cost and the CO$_2$ emission, the kaolinite will be used as such and thus not be dehydroxylated. This is in contrast to what is normally done for geopolymers were a thermally activated aluminosilicate such as metakaolinite or fly ash is used[12].

The research in this paper is in the framework of a water harvesting project with Jordan, moreover to find out which local raw materials can be used (reactive and filler)[13] as a substitute for Portland cement for instance to produce bricks or tiles.

In this paper the choice of the raw material will be discussed. The reactivity of three materials towards NaOH solutions is tested via a dissolution process. The amount of Si and Al in solution is monitored and the solid residues are characterized. From these raw materials, Hiswa kaolinite is chosen to prepare geopolymer samples. The mechanical and thermal properties of the resulting geopolymer will be investigated.

Another part of this research is presented in the work of H. Khoury et al, entitled 'The Effect of Addition of Pozzolanic Tuff on Geopolymers' at this same conference.

EXPERIMENTAL

Materials

The different Jordanian raw materials are: Jordanian Hiswa kaolinite, smectite rich clay, and volcanic tuff. To check the reactivity the samples were sieved to <63 μm. The materials used for fabrication of the geopolymer specimens are: Jordanian Hiswa kaolinite (JHK), Jordanian smectic clay, Jordanian volcanic tuff and Jordanian silica sand as a filler material (JSS), NaOH pellets from Merck with 99 % purity to prepare the alkaline solution with distilled water. Preparation of the geopolymer specimens was done as follows. Kaolinite and sand were mixed using a laboratory mixer. The NaOH solution was added gradually to the kaolinite/sand mixture and mixed for 10 minutes to get an optimum homogeneity. The mixture was divided into 3 stainless steel cylinder moulds (50 height x 25 mm diameter) and compacted with a pressure of 16 MPa by using a hydraulic compressor then de-moulded and weighed. The specimens were cured in a ventilated oven at 80 ˚C for 24 hours. Four series were prepared using different mass ratios of NaOH (8, 12, 16 and 20) and 100 kaolinite, 50 sand, and 22 H_2O, parts in mass.

Techniques

The extent of dissolution of all materials in an alkaline medium was determined by mixing 5 g of each material with 200 ml of a 10 M NaOH solution at room temperature (25 °C). After 5, 24 and 168 h the material was filtered. An Atomic Absorption spectrophotometer (AAS) AAS Analyst 100 from PerkinElmer was used to determine the Al and Si concentrations from the filtered solutions. The solid residue was studied by TGA.

The compressive strength of the geopolymer specimens was measured by using a CONTROLS testing machine (Model T106 modified to suit standard testing), where the load was applied and increased at a displacement rate of 3 mm/min. An average of three specimens per measurement was taken.

The reactivity was studied with a Mettler DSC 822e using reusable high pressure stainless steel sample pans. The scan rate was 5°C/min.

For thermogravimetric analysis a TGA Q5000 from TA Instruments was used, purged with N_2. The sample size was about 10mg and the scan rate 10°C/min.

RESULTS AND DISCUSSION

The different raw materials were subjected to a leaching procedure in a 10 M NaOH solution. Fig 1 shows the amount of Si and Al in solution after 7 days. These amounts were related to the total amount of the element in the raw material. It is clear that for the kaolinite and for the smectite rich clay more than 30% of the initial Si and Al went into solution. For the tuff the values were lower, especially for Al. This indicates a lower reactivity. The kaolinite and smectite rich clays mainly contained kaolinite and quartz[14]. Since quartz is much more stable than kaolinite, it is obvious that the Si in the solution originated from the break-down of kaolinite.

Fig. 2 shows the weight loss of the Jordanian Hiswa kaolinite till 1000°C. A small amount of adsorbed water is present (about 2wt%). The step at about 500°C is due to the dehydroxylation of kaolinite. Comparing the height of this step (9.5wt% between 300 and 800°C) to the weight loss found for pure kaolinite in the same temperature interval (13.76%[12]) shows that the Jordanian Hiswa kaolinite contained about 69wt% of kaolinite by weight. If the amount of Si leached out is related only to the kaolinite, then the amount of kaolinite dissolved would be approximately 45wt%. Thus according to this leaching test the residual sample would still contain 24wt% of kaolinite after 7 days in 10M NaOH solution.

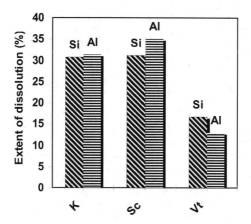

Fig. 1 Extent of dissolution of Si and Al in solution after leaching of kaolinite (K), smectite rich clay (Sc) and volcanic tuff (Vt) in 10M NaOH for 7 days at 25°C

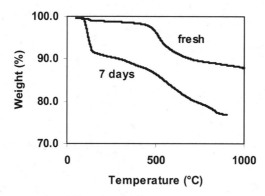

Fig.2 TGA thermograms of Jordanian Hiswa kaolinite before (top, fresh) and after leaching (bottom) for 7 days in 10M NaOH

The weight loss of the residue after the leaching consists of two well separated steps. The first step at about 100°C is due to free water. The second step consists of some overlapping phenomena. The first starts around 300°C and counts for 2wt%. It is probably due to new phases formed. The next step lies in the temperature ranges of the dehydroxylation of kaolinite, although a tail at the high temperature side is remarked. If this step were only due to the dehydroxylation of remaining kaolinite, this amount can be calculated. The height of the step is 8.8%. The remaining kaolinite is therefore estimated to be more than 60%. This value is much higher than expected, since according to the leaching test only 24wt% of kaolinite remains. This weight loss step is thus also due to other new phases than kaolinite.

The fact that new phases are formed during the leaching process implies that the concentrations measured in the solution (Fig.1) do not give a complete picture of the chemical break down of the mineral. The concentration in the solution will be lowered by the precipitation of the new phases. The nature of these phases is elaborated elsewhere[14] and also in the work of H. Khoury et al, 'The Effect of Addition of Pozzolanic Tuff on Geopolymers' at this same conference. As a result it can be stated that some zeolite-like materials such as hydroxysodalite, Na-phillipsite and natrolite are formed in an amorphous matrix.

The leaching out of Si and Al is the first step of the geopolymerisation[15,16]. At least the first part of the DSC thermogram (Fig. 3) is thus due to this process. The height of the DSC signal reflects the reaction rate (heat released per unit of time). The smaller reactivity of the tuff is clear from these data. The tuff only starts to react visibly above 100°C where the kaolinite (JHK) already reacts from 80°C. The smectite-rich clay also starts reacting at 80°C but shows a smaller reactivity below 100°C.

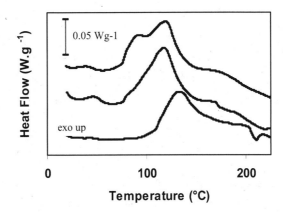

Fig. 3 DSC thermograms of tuff, Jordanian Hiswa kaolinite and Jordanian smectite rich clay with 10 M NaOH

It can also be remarked that the reactivity of these clays is much smaller compared to the reactivity of thermally treated clays such as metakaolinite or fly ash[12]. DSC can also be used to determine the optimum mixing ratio between the two components as was done in previous work[7]. However, this maximum in enthalpy does not necessarily correspond to the maximum mechanical strength (Fig. 4).

Fig. 4 indicates that an optimum compressive strength is obtained for an amount of 16 parts of NaOH to 100 parts of kaolinite by weight. Fig. 5 shows the TGA analysis of this sample. After the loss of some absorbed water, a step between 300 and 700°C, in the neighborhood of the dehydroxylation of kaolinite, is observed. This step counts for 4wt%, which can be related to 30wt% kaolinite that is remaining in the sample after reaction. It is however not sure that this step is solely due to kaolinite, since from the previous results (Fig.2) it was clear that other new phases with weight loss in the same temperature range can be formed.

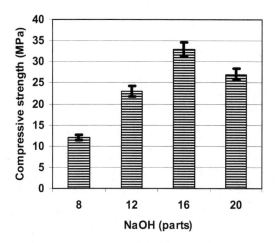

Fig. 4 Compressive strength of geopolymers with different amounts of NaOH (in parts/100 parts of Hiswa kaolinite).

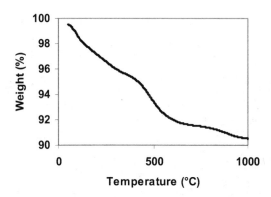

Fig. 5 TGA/DTA thermogram of a geopolymer sample (16 NaOH, 22 Water, 100 JHK)

CONCLUSIONS

The dissolution procedure of minerals in an alkaline solution can give an indication of the reactivity of the minerals. However, since precipitation occurs, even if only 25g of mineral in 1L of solution is used, the concentrations in the solution do not tell exactly how much Si or Al has been leached out. In the case of dissolution of kaolinite containing minerals, it is clear that precipitation of hydrated phases occurs. The kaolinite- and smectite-rich clays dissolve at a comparable rate. The tuff sample dissolves slower. This order is confirmed by DSC. DSC also shows that the reaction starts at about 80°C for the kaolinite- and smectite-rich clay. Since the reactivity at room temperature of the kaolinite/NaOH system is rather low, the materials need to be produced at a higher temperature. This is in contrast to geopolymers produced from metakaolinite, which set in a few hours time at room temperature. For this reason the specimens were cured at 80°C for 24h. The optimum compression strength was found for a mixing ratio of 16g NaOH/100g kaolinite. The mechanical properties of the geopolymers formed from the Jordanian Hiswa Kaolinite are good enough for the construction of water ponds but also for other small constructions.

ACKNOWLEDGEMENT

The financial support of the Deanship of Scientific Research, University of Jordan of the project "Chemical stabilization of natural geomaterials for construction and industrial applications" is highly appreciated. The support of the Flemish (Belgium) Vlaamse Interuniversitaire Raad (VLIR, Contract ZEIN2006PR333) within the "Own Initiatives" program is gratefully acknowledged.

REFERENCES

S. Caijun. et al. in Alkali-activated Cements and Concretes. Taylor & Francis, 2006

[2]J. Gogo, in Geological and Geotechnical Evaluation of Latosols from Ghana, and their Improvement for Construction, Ph.D. Thesis, Vrije Universiteit Brussel, 1990.

[3]M. AlShaaer, H. Cuypers, and J. Wastiels, "Stabilisation of kaolinitic soil for construction purposes by using mineral polymerisation technique", in Proceedings of the 6th International Conference Technology for Developing Countries, (3), Jordan, Hole Musa Resheidat, 1085-1092. (2002)

[4]J. Temuujin, R.P. Williams, A. van Riessen, Effect of mechanical activation of fly ash on the properties of geopolymer cured at ambient temperature, J. Mater. Processing Tech.; 209, 5276-80 (2009)

[5]S.J. O'Connor, K.J.D.A. MacKenzie, New hydroxide-based synthesis method for inorganic polymers. J. Mater. Sci.; 45, 3284-3288 (2010)

[6]P.V. Krivenko, G.Y. Kovalchuk, Directed synthesis of alkaline aluminosilicate minerals in a geocement matrix, J. Mater. Sci.; 42, 2944-2952 (2007)

[7]H. Rahier, B.Van Mele, M. Biesemans, J. Wastiels, X. Wu, J. Mater. Sci.; 31, 71-79. (1996)

[8]H. Rahier, B. Van Mele, J. Wastiels, J. Mater. Sci.; 31, 80 (1996)

[9]Dombrowski, K.; Weil, M.; Buchwald, A., Geopolymer binders. Zkg Internat. 6, 70-80 (2008)

[10]Khoury H in Clays and clay minerals in Jordan. Publications of the University of Jordan, Amman. 2002.

[11]Khoury H. in Industrial rocks and minerals in Jordan. 2nd ed Publications of the University of Jordan, Amman, 2006

[12]Rahier H., Wullaert B., Van Mele B., Influence of the degree of dehydroxylation of kaolinite on the production and structure of low-temperature synthesized aluminosilicate glasses, J. Therm. Anal. Cal., 62, 417 (2000)

[13]Khoury, H., Al Houdali, H., Mubarak, Y., Al Faqir, N., Hanayneh, B., and Esaifan, M. in Mineral Polymerization of Some Industrial Rocks and Minerals in Jordan. Published by the Deanship of Scientific Research, University of Jordan; 1998

[14]Aldabsheh I., Khoury H., Wastiels J., Rahier H., Dissolution Behavior of Jordanian Clay-rich Materials in Alkaline Solutions for Geopolymerization Purpose. Part I, submitted to Applied Clay Minerals

[5]Rahier H., Wastiels J., Biesemans M., Willem R., Van Assche G., Van Mele B., Reaction mechanism, kinetics and high temperature transformations of geopolymers, *J. Mater. Sci.* **42**, 2982 – 2996 (2007)

[6]Hajimohammadi, A.; Provis, J.L.; van Deventer, J.S.J. Effect of alumina release rate on the mechanism of geopolymer gel formation. *Chem Mater.* **22**, 5199-5208 (2008)

PHOSPHATE GEOPOLYMERS

Arun S. Wagh[*]

Inorganic Polymer Solutions, Inc.

4 Helens Way Court,

Naperville, IL 60565

ABSTRACT

Chemically bonded phosphate ceramics/cements (Ceramicrete[TM] for Mg-phosphate compositions) are room temperature-setting inorganic materials with high strength, low porosity, and good durability. These characteristics are similar to alumino-silicate geopolymers. Studying their structure, this author speculated in the annual meeting of the American Ceramic Society in 2004 that they may be considered as a class of geopolymers. Subsequently, Joseph Davidovits in his book on Geopolymers (2008) proposed chemistry and structure of phosphate and phospho-silicate geopolymers based on compositions developed in Argonne. His analysis gives key understanding of the amorphous phase found in these materials that is responsible for their high strength and dense structure. We propose here that the microstructure of phospho-silicate geoplymers consists of a network of crystalline phosphate minerals connected by phospho-silicate geopolymeric amorphous materials. Evidence is provided with X-ray diffraction studies, microstructural studies and strength properties on the presence of amorphous binding phase. We present evidence of such phosphate geopolymers from archeology on the passivation layer on a wrought iron Delhi pillar, which has shown no sign of corrosion despite its antiquity of 1600 years. We also propose for further research that because of more than normal abundance of phosphates in the Nile delta, possibly phospho-silicates might have played a minor binding role in the blocks used in the Egyptian pyramids.

INTRODUCTION

Chemically bonded phosphate ceramics are room or warm-temperature-setting [1-3] ceramics that were developed in Argonne National Laboratory. They are produced by reaction of a sparsely soluble oxide or an oxide mineral and an acid phosphate.

In an earlier presentation in one of the American Ceramic Society meeting [4], we proposed that these ceramics may be considered as phosphate geopolymers. After some initial hesitancy over introducing ortho phosphates as the building blocks of a geopolymer that were considered to be made of tetrahedral silicates and aluminates only, the concept was accepted by none other than Davidovits [5] in his book, *Geopolymer Chemistry and Applications*, by including orthophosphates generalized geopolymers. We discuss the similarities and the difference between the conventional and phosphate geopolymers (C-geopolymers and P-geopolymers) by investigating the structure and synthetic routes employed in forming them, provide some examples of how P-geopolymers existed in ancient India and possibly in Egypt, and how the art of synthesizing them has been reinvented in our work.

GENERALIZED GEOPOLYMERIC STRUCTURES

Figure 1 illustrates the basis of generalized geopolymeric structures, which are tetrahedra of silicon and phosphorous formed with oxygen bonds. Silicon is surrounded by oxygen for charge neutralization, but the net charge of the tetrahedron unit is -1. The same on phosphate tetrahedron is -3/4. Otherwise the two structures are similar, in that they form long chains to form inorganic polymers and hence may be considered as components of generalized geopolymers.

As illustrated by Davidovits [4], the building blocks of C-geopolymers are silicates. To these we now add phosphates. The two fundamental structures are shown in Figure 1.

We use these building blocks to produce various structures as presented in Figure 2, in which the building blocks of silicon and aluminum connect to form siloxo and sialate structures. In a similar manner, phosphorous can form phosphate structures. These three basic structures can connect to form various chains of geopolymers. Just the way siloxo and sialate groups connect to form various C-geopolymers, phosphate also forms phospho-siloxo or phospho-sialate chains of P-geopolymers .

From Table 1, one may infer that the best formulations of P-geopolymers are when they are formed with oxides and phosphate-siloxo or phosphate sialate structures. Such compositions mimic natural minerals of phosphates such as hydroxyapatite, which also contain aluminates or silicates in natural minerals. The X-ray diffraction pattern shown in Figure 3 is a good example of this, in which there is not only the basic crystal structure of magnesium potassium phosphate, but also there exist hydroxyapatite and calcium hydrophosphate and calcium silicate phases. This is also true with alumino silicate based geopolymers that mimic natural durable minerals.

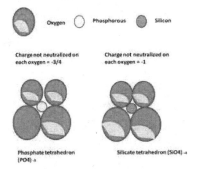

Figure 1. Basic building blocks of phosphate and silicate geopolymers

It should be noted that the P-geopolymers are not limited to these reactions only, but in most cases, they are formed using an oxide or an oxide mineral. For example, the most common compound used as the binders are either magnesium potassium phosphate or calcium phosphate, in which magnesium or calcium oxides are used as the precursors. As we shall see in detail later, these generally tend to give crystalline structures, but when a source of amorphous silica such as fly ash or wollastonite ($CaSiO_3$) is added, amorphous (or glassy or structures with short range disordered) phases are formed in them. This may be seen from the X-ray diffraction patterns in Figure 3. Thus, as in organic polymers, if we were to develop inorganic polymers as the materials containing glassy or phases with short range disorders, the use of amorphous silica is a key ingredient. There exist other methods of producing such short range disorders, which are proprietary. One commercial product that uses these amorphous structures will be described in Section VI in connection with phosphate geopolymeric coatings. Table 1 summarizes some of the most common P-geopolymers commercially used.

DIFFERENCE IN ROUTES OF SYNTHESES OF P- AND C-GEOPOLYMERS

In spite of the structural similarities discussed above, there exists a major difference between the syntheses of P- and C-geopolymers. In both, aqueous solubility of the components is the key to the syntheses. It is necessary to have some dissolved components in the solution for an aqueous based reaction.

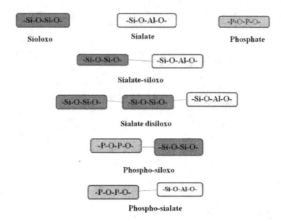

Figure 2. Generalized geopolymeric structures including phosphate geopolymers

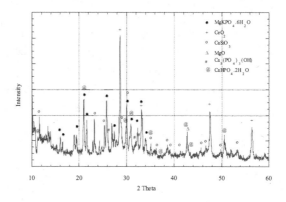

Figure 3: X-ray diffraction pattern of root canal cement developed in
Argonne using wollastonite as the source of chemically
available silica. One may notice amorphous hump indicating

polymeric or amorphous structure and also presence of
hydroxyapatite in the product.

The role of the solubility may be better understood from Figure 4. Silica, which is the major
component of C-geopolymers, has a high solubility only in an extreme alkaline region and its
solubility is very poor in the pH range of 0-10. This requires that all syntheses be carried out in the
alkaline region. Distinct from this, P-geopolymers are synthesized by an acid-base reaction
between an acid phosphate in the pH range of 3-5 with a sparsely soluble oxide or oxide mineral in
the range of $8 - 11$, thus producing a product in the neutral pH range of $7 - 9$. This difference has
several practical implications. With a wide range of oxides with sparse solubility in the near
neutral pH, it has been possible to produce a range of formulations using almost all divalent metal
oxides such as MgO, CuO, CdO, FeO and so on. When the solubility is too low in the near neutral
range as in Al_2O_3, we have been able to heat the mixture slightly, enhance the solubility and
produce ceramics. In the case of the most insoluble iron oxide (Fe_2O_3, hematite), we have been
able to chemically reduce it slightly and produce Fe_3O_4 during the reaction and then produce its
geopolymer. The same is the case with some of the sparsely soluble minerals such as calcium
silicate ($CaSiO_3$, wollastonite) or silicate minerals in fly ash. This freedom has allowed us to
stabilize a range of contaminants from hazardous and radioactive waste streams to pass acceptable
criteria, and to produce several products that are environmentally safe due to their near neutral pH
(or slightly alkaline pH that mimics nature). Details may be found in reference [5].

POLYMERIC STRUCTURES AND PROPERTIES OF P-GEOPOLYMERS

When we first synthesized P-geopolymers with MgO reacting with KH_2PO_4, we obtained a
highly crystalline structure as may be seen in Figure 5(a) that had a compressive strength of
approximately 3000 psi and a porosity of 10-15%. However, once we added a source of
amorphous or chemically available source of silica such as Class C fly ash or wollastonite, the
product was fully dense and we could obtain a compressive strength of 12,000 psi. Such a product
came to be known as "Ceramicrete[TM]." Figure 5(b) shows the amorphous (glassy) phase binding
the cenosphere of fly ash. This glassy phase is also responsible for the broad hump seen in the X-
ray diffraction pattern shown in Figure 3 for a sample produced with addition of wollastonite.
This glassy or amorphous phase provides the polymeric (non crystalline) structure to P-

geopolymers as much as silicate phases give the polymeric structure to C-geopolymers.

Table 1. Some examples of phosphate geopolymers, their uses and structures [6]

Major compound	Applications	Structure and phases
Magnesium potassium phosphate	Radioactive waste immobilization, Binder in Ceramicrete products.	Highly crystalline (see Figure 3). $MgKPO_4 \cdot 6H_2O$ with excess MgO.
Berlinite ceramic	High temperature applications	Crystalline alumina bonded by $AlPO_4$ binder
Iron phosphate	Possible applications are corrosion protection, gamma ray shielding	Highly amorphous, bonded by $FeHPO_4 \cdot nH_2O$, and $Fe_2(HPO_4)_3$
Calcium phosphate	Natural apatite, Additive in dental cement, One of the major components in Ceramicrete	Forms amorphous phases with aluminates and silicates
Zinc phosphate	Dental cement. Very effective with silicates and aluminates	$Zn_3(PO_4)_2 \cdot 4H_2O$ as the main phase
Ceramicrete^{TM}	Rapid-setting cements, Oil well cements, Road repair materials, neutron and gamma ray shields, Radioactive waste immobilization	Magnesium and calcium phosphor silicate amorphous structure. Dense and extremely hard.

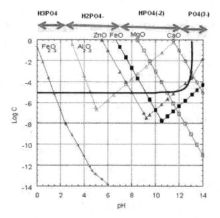

Figure 4. Solubility characteristics of different oxides and
acidity/alkalinity of phosphate ions

This polymeric structure is the most important of the structural aspects of P-geopolymers. It gives better compressive and flexural strength, superior durability and hence has found a wide range of applications. As an example, we will describe its role in the development of paints and coatings in the next Section.

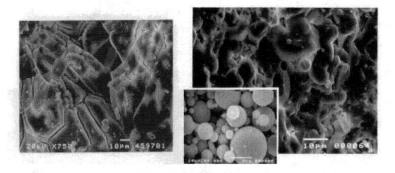

Figure 5 : SEM micrographs of magnesium potassium phosphate and its analog when fly ash is added. Without fly ash, the structure is mainly crystalline, but fly ash provides amorphous silica to the reaction, which converts it into a dense geopolymer.

TWO EXAMPLES FROM ARCHEOLOGY

As Davidovits followed the trail of archeology to discover that the geopolymers were used by ancient Egyptians in building pyramids, and he learnt more about geopolymers from studying the structure of the rocks that make the pyramids, so too we tracked the ancient history of India to understand some of the mysteries that were unresolved till now.

Here we discuss two examples, one the case of a rust proof Delhi iron pillar and the other that of pyramids by revisiting the most recent work on their mystery. While the first case guided us in developing a corrosion protection coating, which we will discuss in the next section, the second one only adds on to what has been confirmed about the geopolymeric structure of rocks in pyramids.

The Delhi iron pillar

The Delhi iron pillar, shown in Figure 6 is one of the scientific mysteries in India. It is a 1600 year old pillar made of wrought iron, which has developed a passivation film of corrosion resistant material by which it does not show any deterioration to the extent that the inscriptions on it are intact.

Figure 6. Delhi iron pillar and inscriptions on it

Dating back to the 4th century A.D., the pillar bears an inscription which states that it was erected as a flagstaff in honor of the Hindu god, Vishnu, and in the memory of the Gupta King Chandragupta II (375-413). 7.3 m tall, with one meter below the ground, the diameter is 48 centimeters at the foot, tapering to 29 cm at the top, just below the base of the wonderfully crafted capital; it weighs approximately 6.5 tonnes, and was manufactured by forged welding [Reference 8].

Recently, Balasubramanyam and Kumar [7] conducted a detailed analysis of this film and determined that the presence of phosphorous and left over slag from the extraction of iron from the iron ore are responsible for the passivation layer. Initial corrosion resulted in the formation of α- and δ-FeOOH. This enhanced phosphorous concentration on the surface. The enhancement of phosphorous on the surface, which catalyzed and formed a compact of amorphous δ-FeOOH next to the surface of the substrate, thereby acting as a passivation layer on the surface of iron. Mossbauer spectroscopy also showed presence of iron phosphate $FePO_4 \cdot 2H_2O$. The golden hue of the pillar when viewed in certain orientations is considered to be due to the presence of iron phosphates.

Using the phosphate geopolymer formulations, we recreated this (or similar) layer in developing a corrosion protection layer for iron, which is described in the next section.

PHOSPHATE GEOPOLYMERS IN PYRAMIDS ?

The next example is that of the pyramids of Giza in Egypt. A controversy over the structure of the blocks that are stacked to construct the pyramids has grown over decades, since ever Davidovits [9] suggested that they were cast using geopolymers . This controversy has been settled for the time being after the detailed analyses of samples of these rocks by Barsoum et al [10].

Barsoum and his colleagues' analysis shows that in addition to high content of calcium, magnesium and silicon, the samples that they analyzed using EDS also contain phosphorous, though inhomogeneously in the range of $0.6 - 1.1$ atom %. They attribute it to the presence of hydroxyapatite ($Ca_5(PO_4)_3 \cdot H_2O$), which in fact is a product of our phosphate geopolymer that is formed during reaction and aging (see Figure 3). Taking the observed atomic percentage of phosphorous in reference 9, we calculated the amount of possible hydroxyapatite in the sample, which amounts to in the range of $3.16 - 5.8$ wt.%, which for a binding phase is significant. In the presence of high amounts of calcium, magnesium, and silica in samples, C-geopolymer is certainly the major phase, but a secondary phase can be P-geopolymer.

PRACTICAL APPLICATION: CORROSION PROTECTION COATINGS

The lesson learnt from the Delhi iron pillar has been most useful in developing corrosion protection coatings during my work with EonCoat, LLC. Using the phosphate geopolymer formulations, we developed corrosion protection coatings both for ambient and high temperature applications. Details of the products may be found on the website www.EonCoat.com, but here we have summarized the science behind it.

EonCoatings are formed by use of oxides of magnesium, aluminum or iron along with a source of silica in the form readily available for reaction (such as fly ash) and reacted with an acid-phosphate in solution. The two components (one mixture of the oxides and the silica source in a wet mixture, and acid solution as the other component) are sprayed plurally on the substrates that need corrosion protection. The two streams mix at the nozzle during spraying and react with the metal substrate to form an intimate bond to protect the substrate from not only ambient atmosphere, but also from saline, acidic, and other chemical environments. High temperature formulations have also been produced to withstand a temperature > 800 °C.

Figure 7, reproduced with permission from theEonCoat^{TM} webpage with permission, shows a comparison between the performance of EonCoat^{TM} and conventional coating. The figure is self-explanatory. It shows superior performance of EonCoat's phosphate geopolymer coating compared to the conventional organic polymer coatings.

Figure 8 is the X-ray diffraction pattern of magnesium phosphate geopolymeric coating on steel. As one may notice, there are very broad peaks in the pattern and also a general broad hump at the base, indicating glassy or amorphous structure of the grains. This is distinct from the magnesium phosphate ceramics produced for structural materials applications that are comparatively sharp. Such an amorphous structure is very much needed for paints and coatings to be flexible and tolerant to stresses produced during aging. In fact, these coatings exhibit excellent wear resistance, strong bonding to the substrates, and can withstand complete and repeated flexural stresses on the substrate.

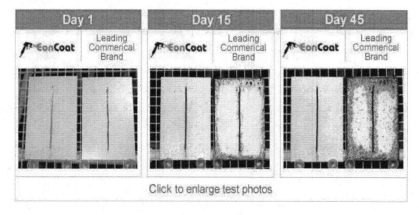

Click to enlarge test photos

Figure 7. Comparison of the rate of rusting with EonCoat™ and conventional paint on steel in salt spray chamber

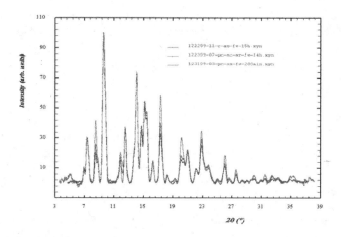

Figure 8. X-ray diffraction pattern of magnesium phosphate coating showing broad peaks and a general hump indicating amorphous (polymeric) structure

The important aspect here is the micro structural similarity between the rust on the Delhi iron pillar and P-geopolymer coatings. As mentioned before, detailed analysis conducted on the Delhi iron pillar rust shows a passivation layer containing FeOOH, and FePO$_4$. We found that EonCoat^{TM} is most effective when the surface of the steel is left slightly rusted to retain oxides of iron prior to spraying. Once sprayed, the oxides react with the phosphate geopolymeric compositions and form two layers. The first layer is rich in FeOOH with traces of FePO$_4$ or FeHPO$_4$ type of minerals and the second layer is that of a fully hardened (or polymerized) dense coating which protects the first layer from the surrounding corrosive atmosphere. Details of this study will be published as a research paper elsewhere. The result is an excellent coating that has shown performance superior to conventional coatings in various tests. Considering the similarities between the phases found on the passivation layer of Delhi iron pillar and the first layer on EonCoat^{TM}, we may anticipate a prolonged protection of steel from corrosion using EonCoat^{TM}.

CONCLUSIONS

The concept of geopolymers may be generalized by introducing bonds between ortho phosphate tetrahedra and those of silicate and aluminates. This broadens the field of geopolymers enormously as may be seen from the work done in Argonne National Laboratory and its applications found by the industry and DOE sites [6].

Phosphate geopolymers are user-friendly and adaptable compared to C-geopolymers. This is because phosphate geopolymers are formed by an acid-base reaction, which is distinct from conventional geopolymers that are formed at very high pH. Thus, selecting a working pH range around neutral pH, it is a lot easier to produce a product instead of using extreme pH. This is the biggest advantage of P-geopolymers even when compared to conventional concrete.

Active research in P-geopolymers has been only during last 17 years and this period is short compared to conventional C-geopolymers. Thus, the basic chemistry and crystallographic studies in P-geopolymers have not been as wide and deep as they have been in C-geopolymers. It is hoped that the basic scientific research in this field will be taken up by the academic community, which eventually may find many new avenues in geopolymer research.

ACKNOWLEDGMENTS

The author thanks the American Ceramic Society for inviting me to present this talk and their partial support, and especially to Prof. Waltraud Kriven for leading this effort. I am also thankful to Anthony Collins, CEO of EonCoat LLC, for his support in this research and permission to reproduce Figures 7 and 8.

REFERENCES

[1] Arun S. Wagh, and S. Y. Jeong, Chemically bonded phosphate ceramics: I, A Dissolution model of formation , *J. Am. Ceram. Soc.*, **86**, 1838-1844 (2003).

[2] Arun S. Wagh, S. Grover, S. Y. Geong, Chemically bonded phosphate ceramics: II, Warm-temperature process for alumina ceramics, *J. Am. Ceram. Soc.*, **86** 1845-1849 (2003).

[3] Arun S. Wagh and S. Y. Jeong, Chemically bonded phosphate ceramics: III, Reduction mecha nism and its application to iron phosphate ceramics, *J. Am. Ceram. Soc.,* **86** 1850-1855 (2003).

[4] A. S. Wagh, Chemically bonded phosphate ceramics - A Novel Class of Geopolymers, A. S. Wagh, *Proc. 106th Ann. Mtg. of the Am. Ceram. Soc.*, Indianapolis, IN, April 18-21, 2004, Published in Ceramic Transactions **165**, ed. J. P. Singh, N. P. Bansal, and W. M. Kriven, 107 – 116 (2004)

[5] Joeseph Davidovits, *Geopolymer Chemistry and Applications*, Geopolymer Institute pub. 592p (2008).

[6] Arun S. Wagh, *Chemically Bonded Phosphate Ceramics*, Elsevier pub. (2004) 300p.

[7] R. Balasubramanyam and A. V. Ramesh Kumar, Characterization of Delhi iron pillar rust by X-ray diffraction, Fourier transform infrared spectroscopy and Mossbauer spectroscopy, *Corrosion Science* **42** 2085-2101 (2000).

[8] R. Balasubramanyam, Iron pillar at Delhi, *Nuevas Investigaciones*, June 10, 2006.

[9] Joseph Davidovits and Margie Morris, *The Pyramids: An Enigma Solved*, Hippocrene Books, New York (1988).

[10] W. Barsoum, A. Ganguly, and G. Hug, Microstructural evidence of reconstituted limestone blocks in the great pyramids of Egypt, J. Am. Ceram. Soc. **89** 3788-3796 (2006).

FOOT NOTE:

*Current address: The author is also with Environmental Sciences Division, Argonne National Laboratory, Argonne, IL 60439, Phone: 630 649 4014, e-mail: asw@anl.gov

Ceramicrete™ is the trademark of Argonne National Laboratory, and EonCoat™ is the trademark of EonCoat, LLC.

Thermal Management Materials and Technologies

3-DIMENSIONAL MODELING OF GRAPHITIC FOAM HEAT SINK

Adrian Bradu and Khairul Alam*
Department of Mechanical Engineering, Ohio University
Athens, OH 45701, USA

ABSTRACT
Recent advances in integrated circuits and aerospace systems have led to the generation of high heat fluxes which require very efficient heat dissipation. Highly conductive graphitic foams are being considered for such heat sink applications because these foams have excellent thermal properties. In this paper, the thermal and flow properties of a graphitic carbon foam is characterized by numerical simulation using the FLUENT software. This analysis is based on a true 3-D model of the foam, which is obtained through x-ray tomography. A solid model of the foam is then developed by reconstructing the surface from the x-ray images. Simulation results with this cell level model of the foam incorporate the effect of the 3-D geometry on the thermo-mechanical properties of the foam.

INTRODUCTION

Graphitic foams can be produced with a wide range of properties, which make them especially suitable for use as thermal management materials. For example, heat sinks can be fabricated based on flow through graphitic foams in a manner similar to the classical heat sinks based on flow over finned surfaces. The advantage of using graphitic foams in such application comes from the fact that the foam possesses a higher surface-to-volume ratio, which leads to a large enhancement in the convective heat flux. Graphitic foams can also be used with phase change materials to provide energy storage.

Open cell graphitic carbon foams were synthesized at the Air Force Research Laboratory (AFRL, Dayton, OH) in 1990's using mesophase pitch precursors. The process includes foaming, followed by carbonization at about 1000°C which produces a low thermal conductivity foam. The final step consists of a graphitization step at temperatures of about 2700°C[1].

Two classical approaches have been used to analyze foams; these are fundamentally different from the true 3-D cell level model that is used in the present study. The first is a macro-approach which consists of averaging the effects of the pores over a representative volume in order to obtain the foam properties. The second approach is a cell level model using an idealized foam model[2]. These two approaches are discussed below.

Bauer[3] and Gibson and Ashby[4] used semi-empirical approaches to define the bulk thermal conductivity on the basis of the relative density of the foam:

$$k_{eff} = \frac{k_b}{k_s} = \left(\frac{\rho_b}{\rho_s}\right)^q \tag{1}$$

Here k_b is the bulk thermal conductivity of the foam, k_s is the ligament thermal conductivity, k_{eff} is the non-dimensional thermal conductivity, ρ_b is the bulk density of the foam, ρ_s is the solid phase density and q is an exponent characterizing the effect of the geometry on the path of heat flow.

Alam & Maruyama[5] used an exponent $1/n$ to represent the Equation (1):

$$k_{eff} = \frac{k_b}{k_s} = \left(\frac{\rho_b}{\rho_s}\right)^{\frac{1}{n}} = \left(1 - \frac{\varphi}{100}\right)^{\frac{1}{n}} \tag{2}$$

107

In the above equation φ represents the porosity of the foam, and n represents the effectiveness of the foam microstructure because the bulk thermal conductivity increases as n increases, the maximum value being achieved for $n = 1$. The exponent n may be called a "density exponent" or "pore conduction factor" and takes values between 0.5 and 1 for majority of foams[2].

Gibson and Ashby[4] determined that most metal foams have the density exponent ranging from 0.555 to 0.625:

$$\left(1 - \frac{\varphi}{100}\right)^{\frac{1}{0.555}} < k_{eff} < \left(1 - \frac{\varphi}{100}\right)^{\frac{1}{0.625}}$$

(3)

Klett et al.[6] found that, for carbon foams there is the need to add the modifier $\beta=0.734$ to Equation (2) in order to account for pore shape.

$$k_{eff} = \frac{k_b}{k_s} = \beta\left(\frac{\rho_b}{\rho_s}\right)^{\frac{1}{n}} = 0.734\left(1 - \frac{\varphi}{100}\right)^{\frac{1}{0.701}}$$

(4)

When the thermal transport take place in both fluid and solid phases the macro-approach of volume averaging has been used, which consists in averaging the heat transfer and fluid flow quantities over a representative elementary volume consisting of both the fluid and solid porous material[7]. Two energy equations are used and the temperatures can be averaged separately for solid and fluid phase[8]. The model is often simplified in order to obtain an analytical solution by making a thermal equilibrium assumption. This macro-approach of volume averaging requires the permeability and the inertial coefficient of the foam as inputs.

An example of this volume averaging is the study by Hunt and Tien[9] to analyze the steady state mass and momentum conservation equations for incompressible flow in porous materials:

$$\nabla \cdot \langle u \rangle = 0$$

(5)

$$\frac{\rho}{\varphi^2}\langle u \cdot \nabla u \rangle = -\nabla \langle p \rangle^f + \frac{\mu}{\varphi}\nabla^2 u - \frac{\mu}{K}\langle u \rangle - \frac{\rho c_f}{\sqrt{K}}|\langle u \rangle|\langle u \rangle$$

(6)

In the above equations $\langle \ \rangle$ represents the volume averaging symbol, u is the fluid velocity vector, p represents the fluid pressure, ρ is the fluid density, μ is the fluid viscosity, φ is the porosity, K is the porous media permeability and c_f is the inertial coefficient.

In order to obtain an analytical solution, Equations (5) and (6) have been simplified by assuming steady state, fully developed flow, without the effect of boundary walls. After applying these approximations, we obtain the Darcy-Forchheimer equation:

$$\frac{\Delta p}{\Delta x} = \frac{\mu}{K}u_D + \frac{\rho c_f}{\sqrt{K}}u_D^2$$

(7)

In Equation (7) u_D is the Darcy velocity:

$$u_D = \frac{\dot{m}}{\rho A_{ch}}$$

(8)

Furthermore, in Equation (7), Δx represents the length of porous media, $\dot{m}$ is the mass flow rate, A_{ch} is the cross-sectional area of the channel that contains the porous media. The "modified" Reynolds number[10] is defined by taking the square root of the permeability as the length scale:

$$\text{Re}_K = \frac{\rho u_D \sqrt{K}}{\mu} \tag{9}$$

Darcy-Forchheimer equation can be used to obtain permeability K and inertial coefficient c_f of the porous media from experimental or numerical simulation results. Beavers and Sparrow[10] and Paek et al.[11] unified the pressure drop properties for various porous materials by introducing the dimensionless friction coefficient as:

$$f = \frac{\frac{\Delta p}{\Delta x} \sqrt{K}}{\rho u_D^2} \tag{10}$$

Beavers and Sparrow[10] combined Equations (7), (9) and (10) to obtain the following:

$$f = c_f + \frac{1}{\text{Re}_K} \tag{11}$$

Paek et al.[11] found that the experimental data for permeability of aluminum foams can be described by:

$$f = 0.105 + \frac{1}{\text{Re}_K} \tag{12}$$

Vafai and Tien[7] found a similar equation for a high porosity foam (Foametal):

$$f = 0.057 + \frac{1}{\text{Re}_K} \tag{13}$$

Several idealized cell level models have been developed by approximating the foam by an idealized geometry in order to reduce the geometrical complexity, and therefore the computational requirements for an accurate representation of the 3-D foam geometry[2]. Metal foams and silicon carbide foams have been approximated as sets of ligaments. However, due to its geometrical complexity, graphitic foam has been represented by many different models. Sihn and Roy[12] proposed a unit cell obtained by subtracting four spheres from a tetrahedron. Druma et al.[13] used a body centered cubic (BCC) structure, and horizontal and vertical ellipses to approximate the carbon foam geometry. Yu et al.[14] constructed the unit cell by subtracting a pore at the center of a cube. Li et al.[15] used a tetrakaidecahedral unit cell for the purpose of representing the carbon foam geometry. These idealized pore-level models are then used to find surface area, porosity, density exponent, fluid flow and thermal transport quantities.

The complex and random geometry of graphitic carbon foams makes it difficult to develop accurate 2-D models, specially for fluid flow. In this study, the analysis is based on a true 3-D cell level model of the foam, which is obtained through x-ray tomography carried out at the Air Force Research Laboratory (AFRL, WPAFB, Ohio). A solid model of the foam is then developed by reconstructing the surfaces from the x-ray images, and the solid geometry is meshed and thermal conductivity is determined by numerical simulations. Simulation of coolant through the foam is also performed using the FLUENT software.

TRUE 3D MODEL RECONSTRUCTION

The 3-D models of two graphitic foams were reconstructed. The first is a foam with the bulk density $\rho_b = 0.37$ g/cm^3 (i.e. porosity $\varphi = 83.2$ %), and the second is a foam with $\rho_b = 0.10$ g/cm^3 (i.e. porosity $\varphi = 95.5$ %). The 3-D digital geometry was reconstructed using images obtained through x-ray tomography carried out at the Air Force Research Laboratory (AFRL, WPAFB, Ohio). Bernsdorf et al.[16], Anghelescu[17], Fiedler et al.[18], and Hugo et al.[19] have used x-ray tomography to obtain 3-D models of foams in order to obtain the thermal and flow characteristics based on the real 3-D microstructure of the foam. This approach has been discussed and compared with the averaging model or a simple pore geometry model approach by Alam et al.[2].

In this approach, the images are first obtained by x-ray tomography; these consist of slices of the physical foam separated in z direction by a voxel size. The voxel size used in the imaging process is 10 μm for each foam sample. During the preparation of the sample for the x-ray tomography, the outside edges of the samples tend to be damaged since some ligaments are crushed as shown in Figure 1(a). Therefore, in order to obtain a 3-D model which will replicate the original geometry of the physical foam with minimum error, only a sub-domain is selected from the middle of the foam for further processing; this sub-domain is highlighted in Figure 1(a) and expanded in Figure 1(b). The domains are segmented by setting a threshold value which will transform the grayscale pictures into binary representation of two phases. In order to obtain a good result, care must be taken to avoid (i) "orphan bubbles" and (ii) "passive islands" (shown in Figure 1(b)) which are either physically implausible, or produce numerical errors.

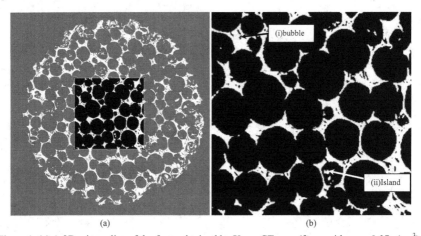

(a)   (b)

Figure 1. (a) A 2D micro-slice of the foam obtained by X-ray CT scan (foam with $\rho_b = 0.37$ g/cm^3); (b) 2D slice after the trimming operation, showing an "(i) orphan bubble, and a (ii) passive island.

The second operation is called labeling and is done using the commercial 3-D data visualization, analysis and modeling software Avizo produced by Visualization Sciences Group (VSG, Burlington, MA). The purpose is to attach labels to pixels in order to identify each subdomain (phase) of the foam. After that, a surface reconstruction operation is carried out. This starts with a triangulation process in which Avizo produces a triangle representation of the boundary between the solid phase and the void using the labels defined in the previous step. Next the software Geomagic Studio[20] is used to fit Non-Uniform Rational B-Spline Surfaces (NURBS) through the points defined by the triangulated surface. The triangulated surface is partitioned into regions defined by high curvature lines, and each of these partitions is further divided into a number of four-sided patches. A grid is attached to each quadrilateral patch and NURBS are fitted using the control points defined by the patches.

The final step consists in obtaining the solid model of the foam and of the fluid in the cells using the NURBS representation of the boundary. This step was done using the commercial finite element pre-processor software HyperMesh[21]. The NURBS are imported into HyperMesh and the solid model of the foam is obtained from the closed shell of surfaces. The solid foam is then inserted into a parallelepiped fluid channel to produce the model of the fluid flowing through the foam cells. HyperMesh is used to generate the mesh for the 3-D model.

THERMAL CONDUCTIVITY NUMERICAL SIMULATION USING THE TRUE 3-D MODEL

Anghelescu[17], Fiedler et al.[18] and Alam et al.[2] have used real 3-D models of foams in order to obtain the bulk thermal conductivity. Anghelescu[17] used the real geometry of a 90% graphitic carbon foam produced at AFRL, and performed finite element analysis to find the bulk thermal conductivity. Fiedler et al.[18] conducted Lattice Monte Carlo simulations using true 3-D models obtained by x-ray tomography to find the thermal conductivity of two aluminum foams.

In order to determine the bulk thermal conductivity, the method presented by Anghelescu[17] is used in this study. The model consists of a graphitic foam placed between two plates, with perfect thermal contact between plates and foam. The top plate is subjected to a heat flux which generates a temperature gradient in the foam that is modeled by the steady state conduction heat transfer equation:

$$\frac{\partial}{\partial x_i}\left(k\frac{\partial T}{\partial x_i}\right) = 0 \qquad i = 1,2,3 \quad \text{(index notation)} \tag{14}$$

In Equation (14), T represents the temperature and k is the thermal conductivity. In order to simplify the model the following assumptions are made: the foam properties do not change with temperature variation, and the effect of the convection and radiation heat transfer on the bulk conductivity is negligible.

The boundary conditions used for setting up the simulations are described as follows:
- The top surface of the upper plate is subject to uniform heat flux;
- The bottom surface of the lower plate is maintained at a constant temperature;
- The contact interface between the plates and the foam is assumed perfect (no thermal contact resistance);
- All other surfaces are considered to be insulated.

The bulk thermal conductivity is found using the temperature gradient given by the numerical simulation in Fourier's law as follows:

$$q_x = -k_b\frac{dT}{dx} = k_b\frac{\Delta T}{\Delta x} \tag{15}$$

In the above equation q_x represents the heat flux, ΔT is the temperature difference between the top and bottom surface of the foam and Δx is the thickness of the graphitic foam. Because the solid phase thermal conductivity k_s is often not known, a non-dimensional bulk thermal conductivity k_{eff} is calculated:

$$k_{eff} = \frac{k_b}{k_s} \tag{16}$$

FLUID FLOW IN THE TRUE 3-D MODEL

The fluid flow simulations at the cell level of true 3-D geometry uses the fluid flow equations and do not require the permeability and inertial coefficients. Bernsdorf et al.[16] used lattice Boltzmann (BGK) automata on a true 3-D representation in order to find the pressure drop through the media. In their case the 3-D computer tomography data can be used directly as the geometry for the lattice Boltzmann simulation, without additional processing. Hugo et al.[19] conducted fluid flow simulation using the Computational Fluid Dynamics software called StarCCM+. They used a true 3-D representation of aluminum foam obtained by x-ray tomography, and the iMorph software to obtain the solid digital model of the foam.

This study uses the method presented by Anghelescu[17] and Alam et al.[2]. The flow characteristics are calculated via direct simulation of flow through a real 3-D digital model in the CFD software Ansys FLUENT[22].

The mass conservation and Navier-Stokes equations for momentum conservation for an incompressible, steady state, laminar flow of a Newtonian fluid are:

$$\frac{\partial u_i}{\partial x_i} = 0 \qquad i = 1, 2, 3 \quad (\text{index notation}) \tag{17}$$

$$\rho u_j \frac{\partial u_i}{\partial x_j} = -\frac{\partial p}{\partial x_i} + \mu \frac{\partial}{\partial x_j}\left(\frac{\partial u_i}{\partial x_j}\right) \qquad i, j = 1, 2, 3 \tag{18}$$

In the above equations, u_i represents the fluid velocity vector, p is the pressure of the fluid, μ is the fluid viscosity and ρ is the fluid density. In order to reduce the computational effort, the fluid properties are considered to be constant and body forces are neglected.

The boundary conditions used for setting up the fluid flow simulations are:
- Inlet with uniform fluid velocity;
- Outlet with constant value of known fluid pressure;
- No-slip condition at the foam-fluid interface;
- Symmetry condition on top, bottom and sides of the model.

RESULTS

The 3-D geometry model of two foams reconstructed using the method presented above are shown in Figures 2-3. These solid models provided good agreement with the measured density provided by the producer of the foam as can be seen from Table I.

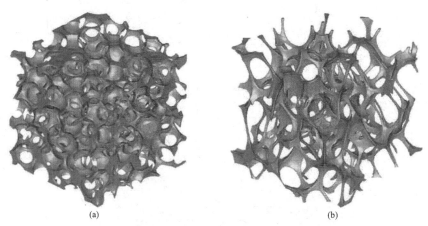

(a)   (b)

Figure 2. (a) 3-D Geometry of the 83.2% porous foam; (b) 3-D Geometry of the 95.5% porous foam (both models are 3 x 3 x 3 mm)

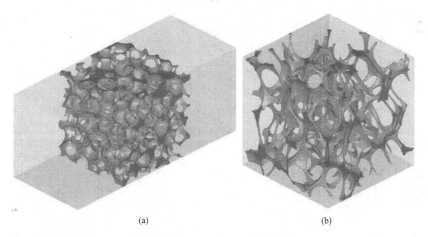

(a)   (b)

Figure 3. (a) 3-D Geometry of the 83.2% porous foam and fluid; (b) 3-D Geometry of the 95.5% porous foam and fluid

Table I. Porosity of the real and reconstructed graphitic carbon foams

Foam sample	Bulk density (g/cm³)	Porosity (%)	
		Actual foam	3-D model
High density sample	0.37	83.2	82.3
Low density sample	0.10	95.5	94.5

The solid models obtained in the previous step were used for thermal and fluid flow simulations. Figure 4 shows the temperature distribution in the 83.2% porous foam, which confirms that, because of the applied boundary conditions, the heat transfer is approximately one dimensional. The results of the thermal analysis for both foams are presented in Table II along with results from other studies. The numerical simulations performed on real geometries yields lower thermal conductivities compared with the results given by simulations based on idealized pores[3]. The values obtained by simulations on real geometries shows good agreement with the experimental results presented by Klett et al.[6]. Fiedler et al.[18] has also shown that the analytical formulas over-predict the numerical and experimental results for thermal conductivities.

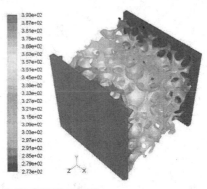

Figure 4. Temperature profile in the 83.2% porous graphitic foam

Table II. Thermal conductivity results

Foam	Source	Type of analysis	k_{eff} (% of solid phase)	n
83. 2 %	Bauer (1993)	Analytical (spherical pores)	9.926	0.77
	Klett et al. (2004)	Experimental	5.780	0.63
	Present analysis	FEM on true foam geometry	5.380	0.61
95.5 %	Bauer (1993)	Analytical (spherical pores)	1.802	0.77
	Klett et al. (2004)	Experimental	0.882	0.65
	Present analysis	FEM on true foam geometry	0.789	0.64

In the next step of the numerical study, fluid flow simulations for inlet velocities between 0.01 m/s and 1.5 m/s were carried out with air flow at 20°C. Figure 5 shows the fluid flows pathlines colored by the velocity magnitudes. The presence of the carbon foam inside the channel decreases the cross-section area of the channel and therefore the fluid velocity increases. As it can be seen from

Figure 5, the maximum velocity increases approximately by 4.2 times in the case of 83.2% porous foam and by 3 times in the case of the 95.5% porous foam,

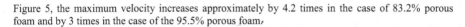
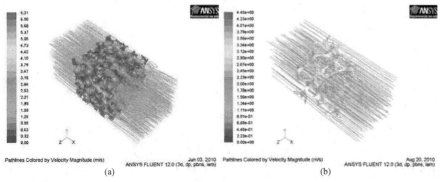

(a)   (b)

Figure 5. (a) Fluid flow pathlines colored by velocity magnitude for an inlet fluid velocity of 1.5 m/s ($\varphi = 83.2\%$); (b) Fluid flow pathlines colored by velocity magnitude for an inlet fluid velocity of 1.5 m/s ($\varphi = 95.5\%$).

The results of the numerical simulations showed that the fluid flow through these two graphitic samples can also be represented by the dimensionless friction coefficient equation (Equation (12)), with the inertial coefficient of about 0.20 to 0.28, which makes the frictional losses higher than the metal foams represented by Equations (12) and (13). The higher pressure drop is expected due to the complex cell geometry of these graphitic foams.

CONCLUSIONS

The 3-D digital models of two graphitic carbon foams with different porosities have been reconstructed from images obtained by x-ray tomography. The reconstructed models are validated by the good agreement between the porosity of the real physical foams and the porosity of the numerical models. The 3-D models were used for analyzing the porous materials in terms of heat transport and fluid flow. In this approach, the transport equations are applied directly to the 3-D reconstructed geometry instead of using an averaged or idealized model.

The simulation values obtained for the thermal conductivity shows good agreement with the experimental data available in literature. The fluid simulations can be used to visualize the flow patterns and to find the flow behavior through the foams. This approach can be used to develop thermal and flow parameters to characterize the foam.

ACKNOWLEDGMENTS
The authors acknowledge the support of Ohio Aerospace Institute (OAI) and the Air Force Research Laboratory (AFRL) for providing funding for this project. The x-ray tomography work was carried out at AFRL. The graphitic foam was supplied by GrafTech. Technical support as well as experimental data was provided by project partners at OAI, GrafTech and TMMT.

FOOTNOTES
*To whom correspondence should be addressed:
Khairul Alam (alam@ohio.edu)
Phone: (740) 593-1558

REFERENCES
[1] M. Brow, R. Watts, M.K. Alam, R. Koch, and K. Lafdi, Characterisation Requirements for Aerospace Thermal Management Applications, *Proceedings of SAMPE 2003*, Dayton, OH (2003).
[2] M. K. Alam, M. S. Anghelescu, and A. Bradu, Computational Model of Porous Media Using True 3-D Images. In A. Öchsner, L. F. M. da Silva & H. Altenbach (Eds.), *Advanced Structured Materials*. Berlin Heidelberg: Springer-Verlag (2010).
[3] T.H. Bauer, A general analytical approach toward the thermal conductivity of porous media, *Int. J. Heat Mass Tran.*, **36(17)**, 4181(1993).
[4] L. J. Gibson, and M. F. Ashby, Cellular solids: Structure and properties (2nd ed.), UK: Cambridge University Press. (1997).
[5] M. K. Alam, and B. Maruyama, Thermal conductivity of graphitic carbon foams, *Exp. Heat Transfer*, **17(3)**, 227-241 (2004).
[6] J. Klett, A. McMillan, N. Gallego, and C. Walls, The role of structure on the thermal properties of graphitic foams, *J. Mater. Sci.*, **39(11)**, 3659-3676 (2004).
[7] K. Vafai, and C. L. Tien, Boundary and inertia effects on flow and heat transfer in porous media, *Int. J. Heat Mass Tran.*, **24(2)**, 195-203 (1981).
[8] V.V. Calmidi, Transport phenomena in high porosity metal foams, Doctoral Dissertation, University of Colorado, Boulder, CO, USA (1998).
[9] M. L. Hunt, and C. L. Tien, Effects of thermal dispersion on forced convection in fibrous media, *Int. J. Heat Mass Tran.*, **31(2)**, 301-309 (1988).
[10] G. S. Beavers, and M. Sparrow, Non-Darcy flow through fibrous porous media, *J. Appl. Mech.-T. ASME*, **36(4)**, 711-714 (1969).
[11] J. W. Paek, B. H. Kang, S. Y. Kim, and J. M. Hyun, Effective thermal conductivity and permeability of aluminum foam materials, *Int. J. Thermophys.*, **21(2)**, 453-464 (2000).
[12] S. Sihn, and A. K. Roy, Modeling and prediction of bulk properties of open-cell carbon foam, *J. Mech. Phys. Solids*, **52(1)**, 167-191 (2004).
[13] C. Druma, M. K. Alam, and A. M. Druma, Finite Element Model of Thermal Transport in Carbon Foams, *J. Sandw. Struct. Mater.*, **6(6)**, 527 (2004).
[14] Q. Yu, B. E. Thompson, and A. G. Straatman, A unit cube-based model for heat transfer and fluid flow in porous carbon foam, *J. Heat Tran.-T. ASME*, **128(4)**, 352-360 (2006).
[15] K. Li, X.-L. Gao, and A. K. Roy, Micromechanical modeling of three-dimensional open-cell foams using the matrix method for spatial frames, *Compos. Part B-Eng.*, **36(3)**, 249-262 (2005).
[16] J. Bernsdorf, G. Brenner, and F. Durst, Numerical Analysis of the Pressure Drop in Porous Media Flow with Lattice Boltzman (BGK) Automata, *Comput. Phys. Commun.*, **129**, 247-255 (2000).
[17] M.S. Anghelescu, Thermal and Mechanical Analysis of Carbon Foam, Doctoral Dissertation, Ohio University, Athens, OH, USA (2008).

[18] T. Fiedler, E. Solorzano, F. Garcia-Moreno, A. Öchsner, I. V. Belova, and G. E. Murch, Lattice Monte Carlo and Experimental Analyses of the Thermal Conductivity of Random-Shaped Cellular Aluminum, *Adv. Eng. Mater.*, **10(11)**, 843-847 (2009).

[19] J.-M. Hugo, B. Brun, F. Topin, and J. Vicente, Conjugate Heat and Mass Transfer in Metal Foams: A Numerical Study for heat Exchanger Design, *Proceedings of the DSL 2009 Conference*, Rome, Italy, 24-26 June (2009).

[20] Geomagic, Inc, Geomagic Studio (12.0.0) [Computer software], Research Triangle Park, NC, USA (2010).

[21] Altair Engineering, HyperMesh (10.0) [Computer software], Troy, MI, USA (2009).

[22] Ansys Fluent Inc., FLUENT (12.0.1) [Computer software], Canonsburg, PA, USA (2009).

ENHANCEMENT OF HEAT CAPACITY OF MOLTEN SALT EUTECTICS USING INORGANIC NANOPARTICLES FOR SOLAR THERMAL ENERGY APPLICATIONS

Donghyun Shin
Department of Mechanical Engineering, Texas A&M University
College Station, TX 77843-3123, USA

Debjyoti Banerjee
Department of Mechanical Engineering, Texas A&M University
College Station, TX 77843-3123, USA
Tel: +1 9798454500, dbanerjee@tamu.edu

ABSTRACT

Thermal energy storage using phase change materials have been widely investigated for concentrating solar power applications. The system efficiency in concentrating solar power applications is affected significantly by the storage temperature, while 70% of the total cost of solar power arises from the material costs of the thermal storage devices. Hence, increasing the operating temperature and decreasing the size of the storage materials can reduce the cost of solar energy. Molten salts (such as carbonate eutectics) have melting points between 200°C and 600°C and are stable up to 600°C. They are also reasonably cheap and environmentally safe. However, the specific heat of the molten salts are low (< 2 J/gK) compared to other conventional thermal storage materials. In this study, silica and magnesia nanoparticles were doped in various carbonate and chloride eutectic mixtures of Ba, Na, Ca, Li and K. Specific heat measurements were performed using a differential scanning calorimeter. The specific heat of the eutectics in both solid and liquid phase were enhanced dramatically (greater than ~ 6% - 20%) on addition of nanoparticles of silica and magnesia, at only 1-1.5% mass concentrations. Materials characterization studies showed that the nanoparticles induced nano-scale phase transformations in the solvent phase.

INTRODUCTION

In concentrating solar power (CSP) systems, solar thermal energy is collected from a large area and focused by mirrors or lenses into a small area in order to obtain thermal energy at high temperature. This high quality thermal energy is then transferred by a heat transfer fluid (HTF) and utilized through thermodynamic cycles, such as Rankine or Stirling cycles, in order to generate electricity. Since the solar energy is available only in the daytime, solar thermal energy storage (TES) is used to store solar thermal energy, which is then utilized during periods of peak demand or for generating electricitywhen solar energy is not available (such as at night or on a cloudy day). Contemporary TES materials include mineral oil, fatty acid, or paraffin wax.[1] These contemporary TES materials are thermally stable up to 400 °C and current CSP, therefore, are limited to operational temperatures that are at 400 °C or lower.

Typically, efficiency of the thermodynamic cycle relies on the temperature difference between the source temperature (such as TES temperature) and the sink temperature (condenser, cooling tower, etc.). Since it is very costly to decrease the condenser temperature, the most practical way to enhance the thermodynamic efficiency is to raise the source temperature (which is ~ 400 °C in contemporary CSP systems). Assuming the condenser temperature is 40 °C, increasing the source temperature from 400 °C to 500 °C can enhance the theoretical Carnot efficiency from 53 % to 60 %. However, very few materials are compatible at such high temperatures.

Alkali-carbonate, alkali-nitrate, alkali-chloride, or eutectic of those are termed as "molten salts" and they are stable up to 700 °C.[2] Using the molten salts as TES material (or HTF) has several benefits: (1) High temperature stability of the molten salts allows CSP to operate higher temperature, which can enhance the thermodynamic efficiency of the system (i.e., reduction in the cost of electricity). (2) Low cost of the molten salts can significantly reduce the cost of electricity. (3) The molten salts are environmentally safe in comparison with other TES materials. This can also reduce the potential cost of disposal of these non-hazardous TES materials. However, these molten salts suffer low thermal conductivity and low specific heat capacity (typically less than 2 J/g°C).[2]

Solid materials doped with nanoparticles are termed as "nanocomposites." Doping minute concentration of nanomaterials into a solid has been frequently reported for anomalously enhanced thermal properties especially with addition of carbon nanotube (CNT). Song and Youn[3] reported 100 % enhanced thermal conductivity of epoxy by adding CNT for 1.5 % concentration by weight. Haggenmueller et al.[4] reported 700 % enhanced thermal conductivity of high density polyethylene by adding single-walled CNT for 20 % concentration by volume. Wang et al.[5] reported 30 % enhancement in thermal conductivity of polymers by adding CNT for 1 % concentration by weight. Haggenmueller et al. proposed that the anomalously enhanced thermal properties of nanocomposites results from the formation of percolation network by interconnected CNTs.[4] Similar to the thermal conductivity of nanocomposites, their specific heat capacity can also be enhanced by doping with nanomaterials. However, few research studies have been performed to explore the change in specific heat capacity of nanocomposites. Also, only a few studies have been reported in the literature for enhanced specific heat capacity of liquids doped with nanoparticles (which are termed as "nanofluids").[6-8]

In this study, we synthesized a new nanocomposite by adding nanoparticles into a molten salt eutectic for TES application in CSP. The new nanocomposite is expected to have higher specific heat capacity, which will significantly reduce the size requirement of TES devices (i.e., reduction in the cost of electricity). SiO_2 nanoparticles were chosen as nanomaterial additives for comparison with previous studies.[6-8] Eutectic of chloride ($BaCl_2$-$CaCl_2$-$LiCl$-$NaCl$ (15.9:34.5:29.1:20.5 by molar ratio) was chosen as the molten salt. The melting point of the particular molten salt was reported to be ~ 378 °C.[9] The specific heat capacity of the nanocomposite was measured by using a differential scanning calorimeter (DSC). Experimentally obtained specific heat capacity was then compared with theoretical specific heat capacity model (This will be discussed later in this study). Transmission electron microscopy (TEM) was employed to measure the size of SiO_2 nanoparticles in the molten salt eutectic as well as to verify minimal agglomeration of nanoparticles during the measurement.

EXPERIMENTAL SETUP

SiO_2 (1% mass concentration) was dispersed into the eutectic of $BaCl_2$-$CaCl_2$-$LiCl$-$NaCl$ (15.9:34.5:29.1:20.5 by molar ratio) to synthesize the nanocomposite by using a two step liquid solution method. SiO_2 nanoparticles were procured from Meliorum Technologies, Inc. $BaCl_2$, $CaCl_2$, $LiCl$, and $NaCl$ were procured from Spectrum Technologies, Inc. The detail procedure of the two step liquid solution method is illustrated in the schematic in Figure 1. The protocol for preparing the pure molten salt eutectic and the nanocomposite are as follows: First, 2.000 mg of SiO_2 nanoparticle, 68.490 mg of $BaCl_2$, 79.206 mg of $CaCl_2$, 25.520 mg of $LiCl$, and 24.784 mg of $NaCl$ were mixed in the dry state. The mass concentration of SiO_2 nanoparticle is thus 1%. The salt eutectic is thus 15.9:34.5:29.1:20.5 by molar ratio. The dry mixture was dissolved in 20 ml of distilled water and the water solution was ultra sonicated by an ultra sonicator (Branson 3510, Branson Ultrasonics Corporation) for ~ 100 minutes to ensure homogeneous dispersion of nanoparticles in molten salt eutectic. The water solution was then placed on a hot plate (C-MAG HP7, IKA) at 150 °C to evaporate water from the solution. After three hours, water was visually evaporated and the hot plate temperature was set to 300 °C for another 2 hours to remove any remaining water molecules, which could be chemically bonded to the nanocomposite. The pure eutectic was also synthesized in the same manner without the addition of SiO_2 nanoparticles.

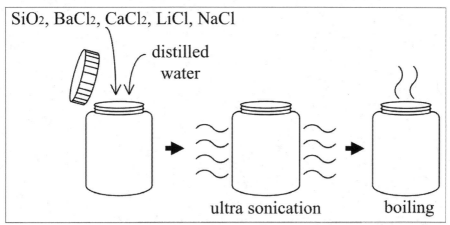

Figure 1. General procedure for preparing the nanocomposite (two step liquid solution method)

The specific heat capacity of the pure eutectic and the nanocomposite were experimentally measured using a differential scanning calorimeter (DSC; Q20, TA Instruments, Inc). Standard aluminum hermetic pan and lid (TA Instruments, Inc) were used to mount a sample in the DSC cell in order to prevent possible weight loss of the sample. The specific heat capacity measurement was performed in accordance with a standard test method (ASTM-E1269). The temperature of the sample was raised at a fixed rate of 20 °C / minute and the heat transfer from the DSC cell to the sample was monitored as a function of temperature. The test was cycled for 4 times to verify the repeatability of measurement. A sapphire standard, whose specific heat capacity is known (1 J/g-°C), and an empty pan - was tested in the same manner. The specific heat capacity of the sample was computed by comparing the differential heat transfer of the sample and the weight of the sample (using the empty pan as the baseline) - to those of the sapphire standard.

The specific heat capacity measured in the experiment was compared with theoretical model (thermal equilibrium model), which is given by:[10]

$$C_{p,t} = \frac{m_{np}C_{p,np} + m_b C_{p,b}}{m_{np} + m_b}$$ (1)

where $C_{p,t}$, $C_{p,np}$, and $C_{p,b}$ are the specific heat capacity of the nanocomposite, the nanoparticle, and the pure eutectic, respectively; while m_{np} is the weight of the nanoparticle and m_b is the weight of the pure eutectic.

RESULT AND DISCUSSION

The specific heat capacity of the pure eutectic ($BaCl_2$-$CaCl_2$-$LiCl$-$NaCl$) and the SiO_2 nanocomposite (at 1% concentration by weight) are shown in Figure 2 and 3. The detailed data from the specific heat capacity measurement are listed in Table 1. The specific heat capacity of the pure eutectic and the nanocomposite (1.0 % mass concentration of nanoparticles) are 0.79 J/g°C and 0.84 J/g°C, respectively. The enhancement in specific heat capacity is 6 ~ 7 % and this value is above the measurement uncertainty (0.5 ~ 2.0 %). Estimate by the theoretical model (Model 1) for the nanocomposite is 0.79 J/g-°C. Since the concentration of SiO$_2$ nanoparticles is extremely small (1.0 %), the prediction by the model do not significantly deviate from the specific heat capacity of the pure eutectic (Figure 2). It implies that the observed enhancement in the specific heat capacity of the nanocomposite cannot be explained by existing macroscopic heat transfer theories.

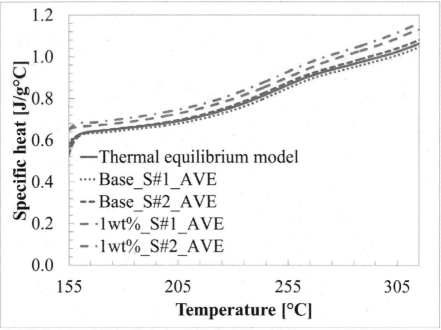

Figure 2. Variation of the specific heat capacity of pure eutectic and nanocomposite with temperature from 155 °C to 315 °C. Theoretical specific heat capacity model is drawn in the figure for comparison.

Possible explanations for the anomalous enhancement observed in these experiments include: (1) Enhancement in the specific heat capacity of nanoparticles contributes to the increase in specific heat capacity of nanocomposite. Literature study shows specific heat capacity of a nanoparticle increases with decrease in the size of nanoparticle. Wang et al.[11] reported specific heat capacity of a nanoparticle can be enhanced by ~25 % in comparison with a bulk solid. Wang et al.[12] theoretically proved that specific heat capacity of a nanoparticle can be increase with decrease in the size of the nanoparticle. (2) The specific surface area (or interfacial area) per unit mass or unit volume of the nanoparticle is enhanced considerably – resulting in high contact area between the nanoparticle and the adhering salt molecules – which can act as additional thermal energy storage mechanism. The interfacial area between a nanoparticle and surrounding salt eutectic is extraordinarily large due to exceptionally large surface area to volume ratio of the nanoparticle. Due to the large increase in the interfacial area, the effect of interfacial interaction of vibrational energies between nanoparticle atoms and the interfacial salt molecules cannot be neglected in nanocomposite and consequently it can act as additional transport mechanism for thermal energy storage.

Moreover, a transmission electron microscopy (TEM; Jeol JEM-2010) was performed in this study. Since nanoparticles tend to agglomerate and precipitate under a certain pH, it is necessary to verify if the nanoparticles are not agglomerated in the nanocomposite. Figure 4 and 5 show SiO_2 nanoparticles in the nanocomposite before and after the specific heat capacity measurement, respectively. It was observed that the nanoparticles did not show any significant agglomeration during the thermal cycling in the DSC.

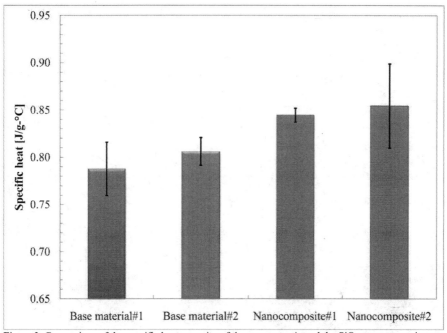

Figure 3. Comparison of the specific heat capacity of the pure eutectic and the SiO_2 nanocomposite.

CONCLUSION

A molten salt eutectic ($BaCl_2$-$CaCl_2$-$LiCl$-$NaCl$) was investigated as a TES material in CSP application. Minute concentration of SiO_2 nanoparticles was doped into the salt eutectic in order to enhance the specific heat capacity. SiO_2 nanoparticles were measured to have a nominal size of ~ 30 nm and the mass concentration of the nanoparticles was fixed at 1%. Two step liquid solution method was employed to synthesize the nanocomposite. Specific heat capacity was measured experimentally by using a differential scanning calorimeter (DSC). The specific heat capacity of the pure eutectic and the nanocomposite was found to be 0.79 J/g°C and 0.84 J/g°C, respectively. The corresponding enhancement in the specific heat capacity is 6~7 % and this value is higher than the measurement uncertainty, 2 %.

Table 1. Average specific heat capacity of the pure eutectic and the nanocomposite (ε: standard deviation).

	Base material#1	Base material#2	Nanocomposite#1	Nanocomposite#2
Repeat#1	0.76	0.79	0.84	0.78
Repeat#2	0.76	0.81	0.84	0.83
Repeat#3	0.77	0.82	0.85	0.85
Repeat#4	0.79	0.82	0.85	0.88
Average	0.77	0.81	0.84	0.83
Enhancement	-	-	7 %	6 %
E	0.01	0.02	0.01	0.04

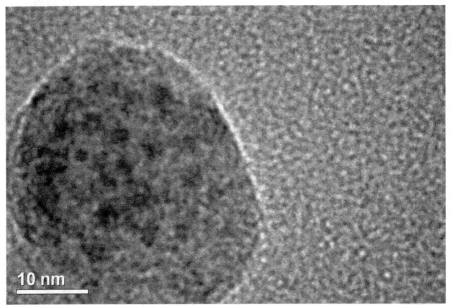

Figure 4. TEM image of SiO_2 nanoparticle in the nanocomposite before repeated thermal cycling in the DSC. The nominal size of SiO_2 nanoparticle is ~ 30 nm.

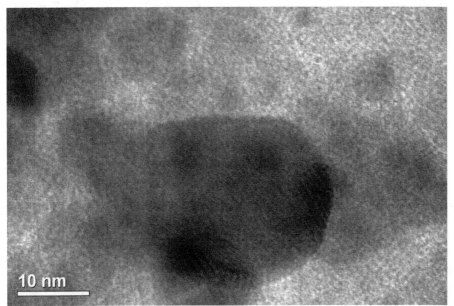

Figure 5. TEM image of SiO_2 nanoparticle in the nanocomposite after repeated thermal cycling in the DSC. The average diameter of SiO_2 nanoparticles is ~30 nm.

The theoretical model (Equation 1) failed to predict the enhancement in the specific heat capacity. Since the mass concentration of SiO_2 nanoparticle was exceptionally small (1%), the analytical estimate for the specific heat capacity of the nanocomposite did not significantly deviate from the specific heat capacity of the pure salt eutectic. This implies that the anomalous enhancement observed in the experimental measurements for the specific heat capacity of the nanocomposite cannot be explained by existing theories. Two independent thermal transport mechanisms were proposed to explain the enhanced specific heat capacity of nanocomposite. (1) Enhanced specific heat capacity of nanoparticles contributes to the enhancement in specific heat capacity of nanocomposite. (2) Additional thermal storage mechanisms were postulated to exist due to the interfacial interaction of the vibration energies between nanoparticle atoms in contact with the salt molecules. Furthermore, transmission electron microscopy (TEM) was employed to observe nanoparticles in the nanocomposite before and after the measurement and no significant agglomeration of nanoparticles was observed. The authors expect that this finding will be beneficial in designing novel thermal energy storage applications in concentrating solar power systems.

ACKNOWLEDGEMENTS

For this study the authors acknowledge the support of the Department of Energy (DOE) Solar Energy Program (Golden, CO) under Grant No. DE-FG36-08GO18154; Amd. M001 (Title: "Molten Salt-Carbon Nanotube Thermal Energy Storage For Concentrating Solar Power Systems").

REFERENCES

1. Farid, M.M., Khudhair, A.M., Razack, S.A., and Al-Hallaj, S., 2004, "A review on phase change energy storage: materials and applications," Energy Conver Manage 45, pp. 1597–1615.

2. Janz, G.J., Allen, C.B., Bansal, N.P., Murphy, R.M., and Tomkins, R.P.T., 1979, "Physical Properties Data Compilations Relevant to Energy Storage, II. Molten Salts: Data on Single and Multi-component Systems," US Govt. Print. Off, Washington, DC

3. Song, Y.S. and Youn, J.R., 2005 "Influence of dispersion states of carbon nanotubes on physical properties of epoxy nanocomposites," Carbon 43, pp. 1378-1385.

4. Haggenmueller, R., Guthy, C., Lukes, J.R., Fischer, J.E., and Winey, K.I., 2007, "Single wall carbon nanotube / polyethylene nanocomposites: thermal and electrical conductivity," Macromolecules 40, pp. 2417-2421.

5. Wang, J., Xie, H., Xin, Z., Li, Y., and Chen, L., 2010, "Enhancing thermal conductivity of palmitic acid based phase change materials with carbon nanotubes as fillers," Solar Energy 84, pp. 339-344.

6. Shin, D. and Banerjee, D., 2010, "Effects of silica nanoparticles on enhancing eutectic carbonate salt specific heat (work in progress)," The International Journal of Structural Changes in Solids 2(2), pp. 25-31.

7. Shin, D. and Banerjee, D., 2011a, "Enhanced specific heat of SiO_2 nanofluid," Journal of Heat Transfer 133(2), pp. 024501.

8. Shin, D. and Banerjee, D., 2011b, "Enhancement of specific heat capacity of high-temperature silica-nanofluids synthesized in alkali chloride salt eutectics for solar thermal-energy storage applications," International Journal of Heat and Mass Transfer (accepted).

9. Janz, G.J., Allen, C.B., Bansal, N.P., Murphy, R.M., and Tomkins, R.P.T., 1978, "Physical properties data compilations relevant to energy storage. I. Molten salts: eutectic data," US Govt. Print. Off, Washington, DC

10. Buongiorno, J., 2006, "Convective transport in nanofluids," ASME J. Heat Transfer 128, pp. 240–250.

11. Wang, L., Tan, Z., Meng, S., Liang, D., and Li, G., 2001, "Enhancement of molar heat capacity of nanostructured Al_2O_3," Journal of Nanoparticle Research 3, pp. 483-487.

12. Wang, B.X., Zhou, L.P., and Peng, X.F., 2006, "Surface and size effects on the specific heat capacity of nanoparticles," Int. J. Thermophys. 27, pp. 139–151.

ENHANCEMENT OF HEAT CAPACITY OF NITRATE SALTS USING MICA NANOPARTICLES

Seunghwan Jung and Debjyoti Banerjee
Mechanical Engineering Department, Texas A&M University,
College Station, TX, United States

ABSTRACT

The specific heat of nitrate salt eutectic (KNO_3: $NaNO_3$ in 60:40 molar ratio) was observed to be enhanced on addition of minute concentration of mica nanoparticles. The measurements were performed using a differential scanning calorimeter (DSC). The specific heat measurements were performed for a temperature range of 150°C-500°C. The melting point of nitrate salt is 220°C. Hence, the specific heat of nitrate salts with low mass concentration of mica nanoparticles was investigated in solid and liquid phase. The experiments were performed for different mass concentrations of mica nanoparticles, which ranged from 0.5% to 2%. The specific heat capacity of the nitrate eutectic in the solid phase was found to be enhanced by as much as 13-15% on addition of the mica nanoparticles. Furthermore, the specific heat values in liquid phase were observed to be enhanced by 13-19%. The implications of these results are in reduced cost of solar thermal energy due to the reduction in material costs for thermal energy storage. Thermal energy storage and the associated material costs are the most dominant factors that contribute to the total cost of solar thermal energy. Hence, application of these mica nanoparticles to nitrate salts can be quite effective in reducing the cost of solar thermal energy since nitrate salt eutectics are typically used as conventional materials for solar thermal energy storage.

INTRODUCTION

Concentrating solar power (CSP) systems are used to generate electricity using the thermal power from insolation. These systems use lenses or mirrors to focus the solar energy collected from a large area to a central receiver. The receiver tube is filled with a fluid which could be oil, molten salt or other materials to capture the incoming thermal radiation. This hot liquid is then used to convert water into steam using a suitable heat exchanger and the generated steam is used to drive a turbine (and generator) for generating electricity. In concentrating solar power systems, one of the advantages is that the heated fluid can be stored economically. In concentrating solar power systems, nitrate molten salts and their eutectics are usually used as materials for Thermal Energy Storage (TES) systems and are also being explored as heat transfer fluid (HTF). If the thermo-physical properties of heat transfer fluid are enhanced, the resulting operational efficiencies can be increased and the operating costs can be reduced in concentrating solar power systems leading to cheaper solar power. One strategy for improvement of the thermo-physical properties is to use nanofluids which are colloidal solvents doped with low mass concentration of nano particles.

Nanofluids are colloidal suspensions of dispersed nanoparticles with very low mass concentration in liquids or solvents [1-4]. A number of literature reports have shown the enhancement of thermo-physical properties of nanofluids compared to those of the pure base fluid (i.e., the neat solvent). Nanofluids were proposed for applications in various heat transfer applications such as in cooling systems [5]. Nanofluids obtained from carbonate and chloride salt eutectics are also being explored for thermal energy storage (TES) for solar thermal power conversion systems [6, 13-24].

A number of literature reports have shown the anomalous enhancement of thermal conductivity of nanofluids through experimental investigation as well as theoretical investigation [1]. However, the study of specific heat of nanofluids has been little reported. Only a limited number of experimental investigations were reported for the measurement of specific heat of aqueous nanofluids. Vajjha and Das (2009) [7] measured the specific heat of three nanofluids containing aluminum oxide (Al_2O_3), zinc oxide (ZnO), and silicon dioxide (SiO_2) nanoparticles in a base fluid such as deionized (DI) water as well as ethylene glycol and DI water mixture (EG/DI water: 60:40). The results have indicated that the specific heat of aqueous nanofluids decreases with an increase in volume concentration of nanoparticles.

However, Nelson et al. (2009) [8] reported an experimental study for oil-based nanofluids where the specific heat of nanofluids was found to be enhanced dramatically. In this study, the poly alpha olefin (PAO) solvent was used as base fluid. Exfoliated graphite nanoparticles (EG) was dispersed in PAO solvent with 0.6% mass concentration. Surprisingly, the specific heat of the PAO-based nanofluid was enhanced by 50% at 0.6% mass concentration of EG nanoparticles. The other thermo-physical properties of the PAO-EG nanofluid were also enhanced dramatically at this minute mass concentration. Forced convective heat transfer experiments were also conducted using the PAO-EG nanofluid. The forced convective heat transfer of the nanofluid was enhanced by ~10% compared to that of the neat PAO solvent for experiments performed using a pin-fin cooler apparatus. Recently, Shin and Banerjee (2009) [6] observed that the specific heat capacity of alkali carbonate eutectic was enhanced by ~25-100% at 1-2% mass concentration of silica (SiO_2) nanoparticles.

In this study, nanofluids were synthesized by dispersing mica nanoparticles in nitrate salt eutectic where consisted of a mixture of potassium nitrate and sodium nitrate (60:40 in molar ratio). This nitrate salt eutectic can be used as HTF and/or TES in CSP. The specific heat values of pure nitrate salt eutectic were measured for a temperature range of 150 °C - 500 °C using a differential scanning calorimeter (DSC). A simple analytical model was developed to predict the specific heat of nanofluids and the predictions from the model were compared with the experimental measurements.

EXPERIMENTAL INVESTIGATION

In this study, potassium nitrate and sodium nitrate mixture (60:40 in molar ratio) was used as the base fluid (neat solvent). This paper presents specific heat measurements of nitrate-based nanofluids containing mica nanoparticles. The measurements were performed using a differential scanning calorimeter (DSC). Measurements were performed over a temperature range of 150°C-500°C. To formulate nitrate-based nanofluids, potassium nitrate and sodium nitrate were procured from Spectrum Inc. The mica nanoparticles also were procured from Spectrum Inc. The nominal size of the mica nanoparticles was ~45 μm (or 325 mesh) according to manufacturer's specification. However, the actual size of the nanoparticles was measured by the authors and was observed by transmission electron microscopy (TEM) images to vary from a few nano-meters to a few micro-meters as shown in Figure 1.

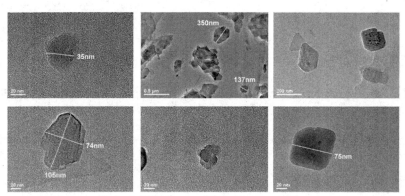

Figure 1. TEM images of the mica nanoparticles

The synthesis protocol for the nitrate-based nanofluids is explained next. Initially, potassium nitrate and sodium nitrate mixture (60:40 in molar ratio) and mica nanoparticles were put into a glass vial of 25ml volume. The total mass of the mixture was 200 mg.The vial was then filled with 20ml of distilled water. To ensure homogenous dispersion of the nanoparticles, the suspension was placed in an ultrasonication bath for 3 hours. The aqueous solution was then placed on a hot plate at 100°C to evaporate the water from the solution. After the evaporation was completed, the nitrate-based eutectic with mica nanoparticles was scraped off. Using this protocol, nitrate-based nanofluids were synthesized with 0.5%, 1%, and 2% mass concentration of mica nanoparticles. These nanofluids were characterized using scanning electron microscopy (SEM) images as shown in Figure 2.

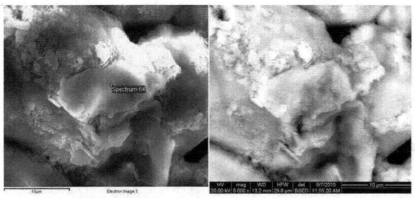

Figure 2. SEM images of the mica/nitrate salt eutectic nanofluid:
(a) Secondary electron image (b) backscattered electron image

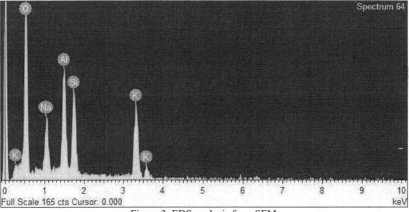

Figure 3. EDS analysis from SEM

SEM images show that the mica nanoparticle was enveloped with nitrate salts. From energy dispersive X-ray spectroscopy (EDS) analysis, the existence of mica nanoparticle was confirmed as shown in Figure 3.

To measure the specific heat of the samples, a standard protocol for differential scanning calorimeter (DSC) testing (ASTM-E1269) was employed. Standard T-zero hermetic pan and lid (TA instruments, Inc) were used to prevent weight loss of samples of nanofluids during repeated thermo-cycling. Before putting into the hermetic pans, the samples were heated at 150°C for 2 hours to remove any chemically adsorbed water molecules. Specific heat values were measured using a differential scanning calorimeter (TA Instrument, Model: Q20). To calculate the specific heat values by the ASTM method, heat flow measurements were performed using an empty pan and the same pan containing a sapphire standard disc, prior to performing the heat flow measurements for the sample (which was placed in the same pan and was hermetically sealed in the final step). The specific heat of the sapphire standard was used as a calibration step in order to calculate the specific heat of the sample of nanofluid for every experiment. A ramping rate of 20°C/min was used during the thermo-cycling - which was repeated 5 times for each sample of nanofluid. The thermo-cycles were repeated to verify if the nanoparticles were stable in the nitrate salt eutectic.

RESULTS AND DISCUSSTION

The measurements of specific heat of the nitrate-based nanofluids were performed for samples with 0.5%, 1%, and 2% mass concentrations of mica nanoparticles. Specific heat of pure nitrate salt eutectic was measured to compare with nitrate-based nanofluids by the ASTM method. Table I shows the measurement result of specific heat of pure nitrate salt eutectic. The temperature range of measurement is from 150°C-500°C. The specific heat value in solid phase is the average value for a temperature range of 160°C-200°C and the specific heat value in liquid phase is the average value for a temperature range of 250°C-495°C. The measurement uncertainty in the pure nitrate salt eutectic is estimated to be 3~4%.

Table I. Measurement of specific heat of pure nitrate salt eutectic

Sample No.	Specific Heat (J/g-K) in solid phase	Specific Heat (J/g-K) in liquid phase
1	1.251	1.408
2	1.098	1.255
3	1.211	1.258
4	1.297	1.471
5	1.153	1.254
6	1.141	1.244
Average	1.191	1.315
STD	0.0746	0.099

The average value of the specific heat of the pure nitrate salt eutectic was measured to be 1.191 J/g-K in the solid phase with a standard deviation of 0.0746J/g-K (6.26%) and 1.315J/g-K in the liquid phase with a standard deviation of 0.099J/g-K (6.13%). The average specific heat values for each thermo-cycle obtained from the DSC measurements for the samples of nanofluids are shown in Tables II, III and IV. These results are also plotted in Figures 4 where the specific heat is plotted as a function of temperature by averaging the measurement results from each thermo-cycle. The measurement uncertainty in the samples of nanofluids is estimated to be 0.13-1.7%.

Table II. Measurement of specific heat of nitrate-based nanofluid with 0.5% mass concentration of mica nanoparticles

Thermo-cycle	Specific Heat (J/g-K) in solid phase	Specific Heat (J/g-K) in liquid phase
1st run	1.384	1.527
2nd run	1.358	1.477
3rd run	1.362	1.477
4th run	1.370	1.479
5th run	1.375	1.483
Average	1.370	1.488
Enhancement	15%	13.2%
STD	0.010	0.021

Table III. Measurement of specific heat of nitrate-based nanofluid with 1% mass concentration of mica nanoparticles

Thermo-cycle	Specific Heat (J/g-K) in solid phase	Specific Heat (J/g-K) in liquid phase
1st run	1.385	1.576
2nd run	1.326	1.502
3rd run	1.324	1.495
4th run	1.318	1.484
5th run	1.314	1.482
Average	1.333	1.508
Enhancement	11.9%	14.7%
STD	0.029	0.039

Table IV. Measurement of specific heat of nitrate-based nanofluid with 2% mass concentration of mica nanoparticles

Thermo-cycle	Specific Heat (J/g-K) in solid phase	Specific Heat (J/g-K) in liquid phase
1st run	1.347	1.556
2nd run	1.342	1.559
3rd run	1.341	1.562
4th run	1.340	1.557
5th run	1.338	1.562
Average	1.342	1.559
Enhancement	12.7%	18.6%
STD	0.003	0.003

The results show that the specific heat values of nitrate-based nanofluids with mica nanoparticles are enhanced compared to pure nitrate salt eutectic for enhancement range of 13-15% in solid phase and 13-19% in liquid phase. In liquid phase, the specific heat values are slightly increased with mass concentration of mica nanoparticles. However, the results indicate that the specific heat values of the nitrate-based nanofluid are less sensitive to the variation of the mass concentration of the mica nanoparticles.

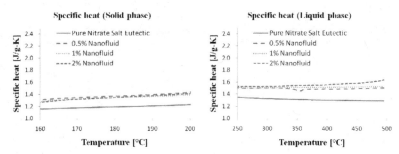

Figure 4. Specific heat values of nitrate-based nanofluid with mica nanoparticles as a function of temperature: (a) specific heat values in solid phase, (b) specific heat values in liquid phase

The mechanism of specific heat enhancement in nitrate-base nanofluid with mica nanoparticles is explained with two aspects. First of all, specific heat values of pure mica nanoparticles are higher than pure nitrate salt eutectic for a temperature range of 150°C-500°C. Figure 5 shows the measurement results of specific heat of pure nitrate salt eutectic and pure mica nanoparticles in solid and liquid phase. Especially, the specific heat values of mica nanoparticles are much higher than the pure nitrate salt eutectic in liquid phase.

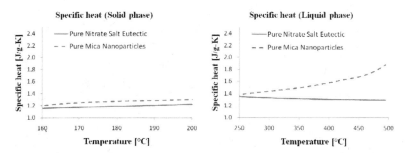

Figure 5. Specific heat values of pure nitrate salt eutectic and mica nanoparticles:
(a) specific heat values in solid phase, (b) specific heat values in liquid phase

Another mechanism of specific heat enhancement is liquid layering at solid-liquid interface. At solid-liquid interfaces semi-crystalline ordering of liquid molecules has been reported in a number of literature reports leading to the formation of a "compressed layer" with a thickness of ~ 1nm. Using a high-resolution transmission electron microscopy (HRTEM) liquid layering was observed at the interface with sapphire by Oh et al. (2005) [9]. This semi-crystalline layer was observed to be adhering to the solid surface as if the liquid molecules were virtually an extension of the underlying crystalline lattice structure and thus had a mass density higher than the liquid phase (Oh et al., 2005) [9]. This layer with a higher density can be termed as the "compressed layer" [10]. A number of reports for molecular dynamics' (MD) simulation have shown that the liquid molecules are observed to form a semi-crystalline ordering of liquid molecules with lower inter-molecular spacing – thereby forming a layer on the surface of the solid with a higher density due to Van der Waals type of inter-molecular interactions. Hence, the liquid molecules adhering to the nanoparticle surface are expected to form a compressed layer in nanofluids. This layer behaves much like a solid phase [11-12].

The existence of compressed layer could be confirmed by measurement of specific heat of nitrate-based nanofluid in solid phase with phase change and without phase change. In the case of phase change, the specific heat was measured for a temperature range of 150°C-500°C. The melting point of nitrate salt eutectic is around 220°C. Thus, phase change is obtained by repeated thermo-cycling. In the case of non-phase change, the specific heat was measured for a temperature range of 150°C-200°C. Because the measurement range is below melting point, the phase change does not occur in this case. Figure 6 indicates that specific heat values in solid phase for each case. The results show that the specific heat values of nanofluids with phase change are higher than those without phase change in solid phase.

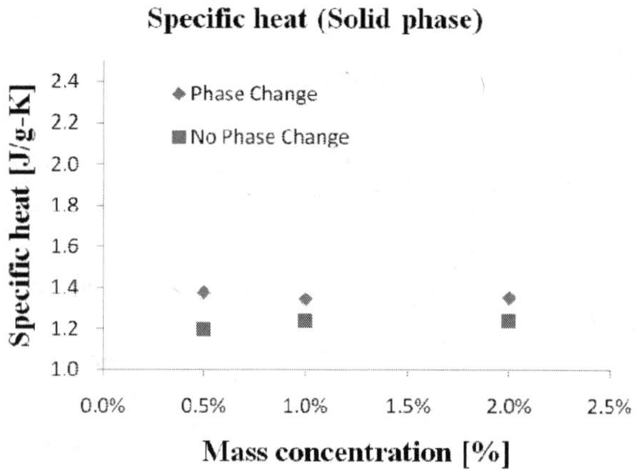

Specific heat (Solid phase)

Figure 6. Specific heat values of nanofluids with mass concentrations in solid phase with phase change and without phase change.

Based on these mechanisms, a simple analytical model for prediction of specific heat values of nanofluids was developed by Jung et al. (2010) [10]. The model considers the contribution to the total specific heat of nanofluid by compressed layer formed at solid-liquid interface. Figure 7 shows the schematic representation of the model. Considering a nanofluid with a total mass of M with a mass concentration of the nanoparticles defined as x, the specific heat of nanofluid was expressed as

$$C_{p,\,total} = \frac{[MxC_{p,\,n}]+[\frac{Mx}{m_n}m_sC_{p,\,s}]+[(M - Mx - \frac{Mx}{m_n}m_s)C_{p,\,l}]}{M} \tag{1}$$

where $C_{p,n}$, $C_{p,s}$ and $C_{p,l}$ are the specific heat of nanoparticle, compressed layer and the liquid phase, respectively. The mass of a spherical nanoparticle of diameter (D_{np}) can be expressed as

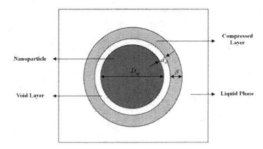

Figure 7. Schematic for the compressed layer along with the void space enveloping a nanoparticle.

$$m_n = \rho_n V_n = \rho_n \left(\frac{\pi D_{np}^3}{6} \right)$$

(2)

where ρ_n and V_n are the density and volume of nanoparticle, respectively. The mass of compressed layer surrounding a single nanoparticle can be expressed as:

$$m_s = \rho_s V_s = \rho_s \frac{4}{3} \pi \left[\left(\frac{D_{np}}{2} + \delta + d_{sl} \right)^3 - \left(\frac{D_{np}}{2} + d_{sl} \right)^3 \right]$$

(3)

where ρ_s and V_s are the density and volume of compressed phase, respectively. Void space (d_{sl}) and thickness of compressed layer (δ) are shown in Figure 7.

To calculate the specific heat value of mica/nitrate salt eutectic nanofluids using the simple analytical model, the thermo-physical properties of the pure nitrate salt eutectic, the compressed layer, and the mica nanoparticles are summarized in Table V.

Table V. The thermo-physical property values of the pure nitrate salt eutectic, the compressed layer, and the mica nanoparticles as well as the thickness of the compressed layer and the void space.

Nanofluid	ρ_n (kg/m³)	ρ_s (kg/m³)	$C_{p,n}$ (J/g-K)	$C_{p,s}$ (J/g-K)	$C_{p,l}$ (J/g-K)	d_{sl} (nm)	δ (nm)
Mica/Nitrate	986	2770.5	1.55	131.5	1.315	0.3	1

The molecules in the compressed layer surrounding the nanoparticle surface are reported to behave like a solid phase [11-12]. Currently there is no available experimental or numerical data for predicting the thermo-physical properties of the compressed layer. In this study, density of compressed layer is assumed to be 50% higher than pure nitrate salt eutectic in liquid phase. The specific heat of compressed layer is assumed to be 100 times of specific heat of the pure nitrate salt eutectic. In a number of molecular dynamics (MD) simulations, the liquid molecules were found to represent a disordered structure observed in the liquid phase at a distance greater than ~1nm from the nanoparticle surface. Based on these results, the thickness of the compressed layer is assumed to be 1nm. The void space between the nanoparticle surface and the first molecule in the compressed layer is estimated to be 0.3nm. The calculation results are compared with the experimental results as shown in Figure 8.

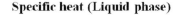

Specific heat (Liquid phase)

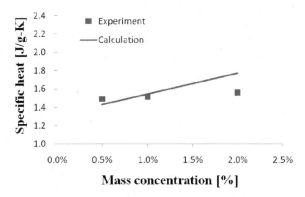

Figure 8. Calculation results of specific heat values and the experimental results

The calculation results from the simple analytical model for the specific heat values of mica/nitrate salt eutectic nanofluids are found to be in close agreement with the experimental results. In the case of nanofluid with 2.0% mass concentration, calculation results are much higher than the experimental result. Based on the analytical model, specific heat value of well-dispersed nanofluid is much higher than one of the nanofluid with agglomerated nanoparticles. It is possible that at higher mass concentrations the interactions between the nanoparticles caused rapid agglomeration and thus lower values of the specific heat (than the values expected from the analytical predictions). Thus, the synthesis protocol needs to be adjusted to obtain perfectly dispersed nanoparticles in order to match the experimental results with the analytical predictions for nanofluids with higher mass concentrations.

SUMMARY/CONCLUSION
Experimental results show that the specific heat values of nitrate-based nanofluids with mica nanoparticles are enhanced by 13-15% in solid phase and 13-19% in liquid phase compared to pure nitrate salt eutectic in 60:40 molar ratio. The enhancement in the specific heat values in the liquid phase increases with the increase of mass concentration. However the specific heat values in the solid phase do not show a similar trend. The main reason for the specific heat enhancement is assumed to be due to the layering of the solvent molecules at solid-liquid interface on the nanoparticles. The evidence for this hypothesis is that from the experimental results - the specific heat values in solid phase of the samples undergoing phase change are much higher than that of the samples without phase change. Furthermore, the specific heat of mica nanoparticles is higher than the pure nitrate salt eutectic in the temperature range of 150°C - 500°C. The calculation results using the simple analytical model for prediction of the specific heat of nanofluids with the effect of compressed layer are in good agreement with the experimental results. The model indicates that well-dispersed nanofluids have higher enhancement of specific heat values compared to nanofluids with nanoparticle agglomeration.

ACKNOWLEDGEMENTS

For this study the authors acknowledge the support of the Department of Energy (DOE) Solar Energy Program (Golden, CO) under Grant No. DE-FG36-08GO18154; Amd. M001 (Title: "Molten Salt-Carbon Nanotube Thermal Energy Storage For Concentrating Solar Power Systems").

REFERENCE

[1] X. Wang and A.S. Majumdar, Heat Transfer Characteristics of Nanofluids: A Review, Int. J. Thermal Sciences, 46, 1–19 (2007).

[2] P. Keblinski, J.A. Eastman, and D.G. Cahill, Nanofluids for thermal transport, Materials Today, 8, 36-44 (2005).

[3] S.K. Das, S.U.S Choi, and H.E. Patel, Heat transfer in nanofluids – a review, Heat Transfer Eng., 27, 3–19 (2006).

[4] W. Yu, D.M. France, J.L. Routbort, and S.U.S Choi, Review and comparison of nanofluid thermal conductivity and heat transfer enhancements, Heat Transfer Eng., 29 (5), 432–460 (2008).

[5] L. Li, Y. Zhang, H. Ma, and M. Yang, Molecular dynamics simulation of effect of liquid layering around the nanoparticle on the enhanced thermal conductivity of nanofluids, J. Nanopart. Res., 12, 811-821 (2009).

[6] D. Shin and D. Banerjee, Investigation of Nanofluids for Solar Thermal Energy Storage Applications, ASME Energy Sustainability Conference, Paper No. ES2009-90465 (2009).

[7] R.S. Vajjha and D.K. Das, Specific Heat Measurement of Three nanofluids and Development of New Correlations, Int. J. Heat Mass Transf., 131, 071601-7(2009).

[8] I.C. Nelson, D. Banerjee, and R. Ponnappan, Flow Loop Experiments Using Polyalphaolefin Nanofluids, J. Thermophys. Heat Transfer, 23, 752-761 (2009).

[9] S.H. Oh, Y. Kgoldffmann, C. Scheu, W.D. Kaplan, and M. Rühle, Ordered liquid aluminum at the interface with sapphire, Science, 310, 661-663 (2005).

[10] S. Jung and D. Banerjee, Experimental Validation of a Simple Analytical Model for Specific Heat Capacity of Aqueous Nanofluids, *SAE Power Systems Conference*, Paper No. 2010-01-0134 (2010).

[11] H. Xie, M. Fujii, and X. Zhang, Effect of interfacial nanolayer on the effective thermal conductivity of nanoparticle-fluid mixture, Int. J. Heat Mass Transf., 48, 2926–2932 (2005).

[12] C.J. Yu, A.G. Richter, A. Datta, M.K. Durbin, and P. Dutta, Molecular layering in a liquid on a solid substrate: an X-ray reflectivity study, Physica B, 283, 27–31 (2000).

[13] Shin, D., and Banerjee, D., "Enhanced Specific Heat of Silica Nanofluid", ASME Journal of Heat Transfer , doi:10.1115/1.4002600, Volume 133, Issue 2, 024501 (4 pages) , February, 2011.

[14] Shin, D., and Banerjee, D., "Enhancement of specific heat of high-temperature silica-nanofluids synthesized in alkali chloride salt eutectics for solar thermal-energy storage applications", International Journal of Heat and Mass Transfer (accepted, in print), doi:10.1016/j.ijheatmasstransfer.2010.11.017, 2011.

[15] Shin, D., and Banerjee, D., "Effects of silica nanoparticles on enhancing the specific heat capacity of carbonate salt eutectic (work in progress)", International Journal of Structural Change in Solids – Mechanics and Applications", Vol. 2, No. 2, pp. 25-31, November, 2010.

Shin, D., and Banerjee, D., "Experimental Investigation of molten salt nanofluid for solar thermal energy application", Paper No. AJTEC2011-44375, ASME/JSME 8th Thermal Engineering Joint Conference, Honolulu, Hawaii, March 13-17, 2011.

16 Shin, D., and Banerjee, D., "Enhanced specific heat capacity of molten salt-metal oxide nanofluid as heat transfer fluid for solar thermal applications", Paper No. SAE-10PSC-0136, SAE Power Systems Conference, November 2-4, Fort Worth, Texas, 2010.

[17]Shin, D., Jo, B., Kwak, H., and Banerjee, D., "Investigation of High Temperature Nanofluids for Solar Thermal Power Conversion and Storage Applications", Paper No. IHTC14-23296, 14th International Heat Transfer Conference, August 8-13, Washington, D.C., 2010.

[18]Shin, D., and Banerjee, D., "Enhanced Thermal Properties of PCM Based Nanofluid For Solar Thermal Energy Storage", Paper No. ES2010-90293, ASME 4th International Conference on Energy Sustainability, May 17-22, Phoenix, AZ, 2010.

[19]Kwak, H., Shin, D., and Banerjee, D., "Enhanced Sensible Heat Capacity of Molten Salt Based Nanofluid For Solar Thermal Energy Storage Application", Paper No. ES2010-90295, ASME 4th International Conference on Energy Sustainability, May 17-22, Phoenix, AZ, 2010.

[20]Jo, B., and Banerjee, D., "Study of High Temperature Nanofluids using Carbon Nanotubes (CNT) for Solar Thermal Storage Applications", Paper No. ES2010-90299, ASME 4th International Conference on Energy Sustainability, May 17-22, Phoenix, AZ, 2010.

[21]Shin, D., and Banerjee, D., "Investigation of Nanofluids for Solar Thermal Storage Applications", Paper No. ES2009-90465, ASME Energy Sustainability Conference, July 20-25, San Francisco, CA, 2009.

[22]Jung, S., Jo, B., Shin, D., and Banerjee, D., "Experimental Validation of a Simple Analytical Model for Specific Heat Capacity of Aqueous Nanofluids", Paper No. SAE- 10PSC-0134, SAE Power Systems Conference, November 2-4, Fort Worth, Texas, 2010.

[23]Jung, S., and Banerjee, D., "A simple analytical model for specific heat of nanofluid with tube shaped and disc shaped nanoparticles", Paper No. AJTEC2011-44372, ASME/JSME 8th Thermal Engineering Joint Conference, Honolulu, Hawaii, March 13-17, 2011

[24]Jo, B., and Banerjee, D., "Interfacial Thermal Resistance between a Carbon Nanoparticle and Molten Salt Eutectic: Effect of material properties, particle shapes and sizes", Paper No. AJTEC2011-44373, ASME/JSME 8th Thermal Engineering Joint Conference, Honolulu, Hawaii, March 13-17, 2011.

ENHANCED VISCOSITY OF AQUEOUS SILICA NANOFLUIDS

Byeongnam Jo and Debjyoti Banerjee
Department of Mechanical Engineering, Texas A&M University
College Station, TX, USA

ABSTRACT

The viscosity of aqueous silica (SiO_2) nanofluids was measured using a parallel-disk rotational rheometer. The silica nanoparticles had a nominal diameter of ~10 nm. The mass concentration of the silica nanoparticles was varied in this study from 0.1% - 20%. Each nanofluid sample was sonicated in an ultra-sonication bath for the same duration. A convection oven within the test section was used to investigate the variation of the viscosity as a function of temperature for the aqueous silica nanofluids. The viscosity measurements were also performed by increasing the shear rate. The pH of the nanofluids was maintained at a fixed value for each sample. The results show that the viscosity of the nanofluids follows a shear thinning behavior since the viscosity decreased as the shear rates were increased. This shear thinning behavior was more acute at higher temperatures. The viscosity of the nanofluids was found to increase with increase in the concentration of the nanoparticles, where the pH was maintained at a constant value.

INTRODUCTION

Solvents doped with a stable suspension of nanoparticles are termed as nanofluids. In these suspensions the nanoparticle size is usually smaller than 100 nm. Conventional fluids such as water and ethylene glycol were employed as the solvent in various studies[1]. Various studies reported anomalous enhancement in the thermal conductivity and diffusivity of nanofluids[2-6]. Pool boiling of nanofluids were also enhanced dramatically[7-10]. For example, critical heat flux (CHF) of aqueous nanofluids was enhanced by up to 3 times – compared to that of pure water[7, 8].

Due to these enhanced thermal properties, nanofluids are considered to be attractive for a wide range of heat transfer applications in electronics, automotive, and energy applications. Furthermore, enhanced characteristics of the nanofluids associated with mass transfer and antibacterial activity makes it attractive for prospective applications like transportation and medical devices. In order to determine the feasibility of using nanofluids for these applications, their rheological behaviors and properties should be characterized. Nevertheless, only few data have been reported in the literature for the viscosity of nanofluids. In 2006, Kang et al. performed viscosity measurements with various nanofluids, including water-silver, water-silica, and ethylene glycol (EG)-diamond nanofluids[11]. They reported that the viscosity was increased by ~30 % in aqueous nanofluids and by ~ 50% for the EG-diamond nanofluid. In 2008, Nguyen et al. performed experiments to examine the effects of temperature and particle volume concentration on the dynamic viscosity using water-alumina nanofluids[12]. The same year another study reported that for aqueous nanofluids obtained by dispersing TiO_2 or Al_2O_3 nanoparticles in water the viscosity was enhanced by ~80%[13]. This was consistent with prior experimental studies reported by Wang and collaborators[14]. Xia et al. performed experimental measurements of the viscosity of EG-based nanofluids and aqueous nanofluids, and they reported that the viscosity enhancements of EG-based nanofluids were smaller than those of aqueous nanofluids[15].

Even though a few experimental measurements for the nanofluid viscosity have been reported, they are not adequate to provide a comprehensive description of the rheological behaviors of nanofluids and to develop proper theoretical models. The aim of the present study is to add to the database for the rheological property data for nanofluids and to provide experimentally measured data for the viscosity of aqueous nanofluids by dispersing silica (SiO_2) nanoparticles with a nominal mean diameter of ~10 nm. The rheological behavior of the aqueous silica nanofluid was recorded using a rotational parallel-disk rheometer. The effect of particle concentrations on the viscosity of the nanofluid was investigated for six different mass concentrations ranging from 0.1% - 20%. In order to maintain homogeneous dispersion of the nanoparticles, pH values of each nanofluid were maintained at a fixed value. Additionally, the viscosity was measured at 4 temperature values ranging between 30 °C and 60 °C.

EXPERIMENTS

Synthesis of nanofluids

Water-based silica (SiO₂) nanofluids samples used in this study were prepared by diluting a commerciall available stock solution of the nanofluids. The stock solution of the aqueous silica nanofluid was available commercially at a mass concentration of 30% (procured from Alfa Aesar) which consisted of spherical silica (SiO₂) nanoparticles with a nominal diameter of ~10 nm. Figure 1 shows the Transmission Electron Microscopy (TEM, JEOL 2010) images of the silica nanoparticles from the stock solution. The TEM images (Figure 1) showed that the diameter of the silica nanoparticles was not exactly 10 nm, but the mean diameter of the particles was measured to be close to 10 nm or were slightly larger. As shown in Figure 1, the nanoparticles were approximately spherical in shape.

Figure 1. TEM images of silica nanoparticles

In order to explore the rheological behavior of the aqueous nanofluids, the nanoparticle concentrations were varied, and the measurements were performed for six different mass concentrations of 0.1%, 1%, 5%, 10%, and 20%. Nanofluid samples for each concentration value was synthesized by diluting the stock solution of the nanofluid. Usually pH values are controlled by introducing additives to obtain stable suspension of the nanoparticles. In fact, the pH value of the stock solution of the nanofluid was set to stabilize the silica nanoparticles in the suspensions. However, on diluting the stock solution the pH values would be changed drastically – especially for the samples containing the low concentration of nanoparticles. Hence, the pH values of diluted nanofluids were maintained at a fixed value to correspond to the same pH as the stock solution. The pH value of the stock solution was measured to be 10.3. However, on dilution of the stock solution the pH value was decreased. For instance, the pH of the nanofluid was measured to be 8.9 for mass concentration of 0.1%. Hence to change the pH of the diluted samples to 10, Ammonium Hydroxide (NH₄OH) was added. However, for the nanofluids with mass concentration of 10% and 20% the pH was found to be almost unchanged at a value of ~10.3 upon dilution; hence ammonium hydroxide additive was not required. The pH variation of the six nanofluids (before using the ammonium hydroxide additive) are summarized in Table 1.

Table 1. Initial and controlled pH values of each nanofluids

Concentration [wt. %]	Initial pH	pH (after addition of ammonium hydroxide)
0.1	8.86	10.21
1	9.69	10.24
5	10.12	10.23
10	10.26	N/A
20	10.32	N/A

Ultrasonication was performed in a sonication bath for 1 hour to ensure homogenous dispersion of the silica nanoparticles. As shown in Figure 2, the silica nanofluids ranged from a faded milky white dispersion which became denser as the mass concentration was increased. Nevertheless, no sedimentation were observed in the vials of all the nanofluid samples – showing that a stable dispersion was obtained. In addition, each sample was sonicated for 30 minutes immediately prior to performing the experiments in the rheometer. This was performed to minimize the effect of nanoparticle agglomeration on the viscosity measurements of the nanofluids.

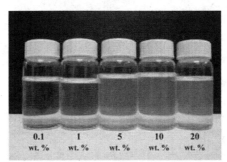

Figure 2. Images of silica nanofluids

Experimental Procedure

A parallel-disk rotational Rheometer (AR-2000ex, TA Instrument) was employed to measure viscosity of the nanofluids. Disposable aluminum plates with 40 mm diameter were used and the gap distance between two plates was 500 μm in this study. Shear viscosity of the nanofluids was measured by steady-state mode (steady rate-sweep test) from 100-1000 [1/s] where the shear rate was increased logarithmically in 10 steps.

First of all, viscosity of pure water (distilled water) was measured between 30 °C and 60 °C in increments of 10 °C to compare with literature values and to verify the accuracy of the experiments. Temperature values were measured from the bottom of a lower plate and was regulated by external convection heaters installed in the rheometer. We varied the gap distances to identify the optimal value of the gap distance. Finally, we decided 500 μm as the optimal value for the gap size in this study.

After completing baseline tests using water, the viscosity of the nanofluids were measured for the same temperature range. The viscosity of the silica nanofluids at the shear rate of 1000 [1/s] was compared with that of pure water (obtained from the literature).

RESULTS

Viscosity of pure water

Figure 3 displays the viscosity behaviors of pure water at four different temperatures. As mentioned above, the viscosity of pure water was determined by averaging measured viscosity between 100 and 1000 [1/s]. As shown in Figure 3, viscosities of pure water measured in the present study decreased with an increase in temperature, which was consistent with data reported in the literature. Moreover, once a 500 μm as a gap distance was employed, the viscosities of pure water were found to be in good agreement over this temperature range. The maximum error was up to 4.3 % at 60 °C.

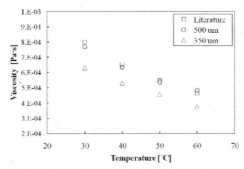

Figure 3. Viscosity of pure water as temperature varies: effect of gap size

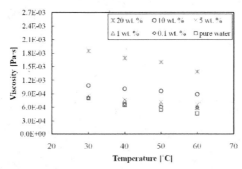

Figure 4. Viscosity of silica nanofluids: effect of particle concentrations

However, it was found that even though the same trend was observed, viscosities of pure water was shifted down when using 350 μm as the gap between two parallel plates (probably due to dominance of the surface tension effects). Overall behavior in the measurements were very similar with the results from 500 μm gap distance, but lower values were obtained with this experimental condition. Therefore, we fixed the gap at 500 μm for the nanofluids experiments.

Viscosity of nanofluids

Viscosities of the silica nanofluids were measured under the same experimental conditions. Figure 4 shows the viscosity of the nanofluids as temperatures and concentrations were varied. In this figure, the viscosity of the nanofluids indicates the value observed at the shear rate of 1000 [1/s] and the literature data for pure water was also plotted. From Figure 4, it is observed that the viscosity of the solvent was enhanced by adding silica nanoparticles. At relatively lower concentrations like 0.1, 1, and 5%, the viscosity was not significantly increased, especially in low temperature. However, the viscosity of the nanofluids was remarkably elevated at nanoparticle concentrations over 5%. At the largest concentration of 20%, the viscosity was enhanced by ~300 %. The measured viscosities of water and the nanofluids are summarized in Table 2.

Table 2. Measured viscosity of water and water-based silica nanofluids

	Viscosity [$\times 10^{-4}$ Pa·s]			
Temperature [°C]	30	40	50	60
Pure water [a] [16]	8.00	6.53	5.46	4.64
Pure water [b]	7.70	6.34	5.36	4.84
0.1 wt. %	8.06	6.72	6.13	6.02
1 wt. %	8.29	7.51	6.82	5.86
5 wt. %	8.61	7.77	7.23	6.66
10 wt. %	10.9	10.1	9.63	8.90
20 wt. %	18.6	17.0	16.1	13.9

[a]: literature values , [b]: measured values

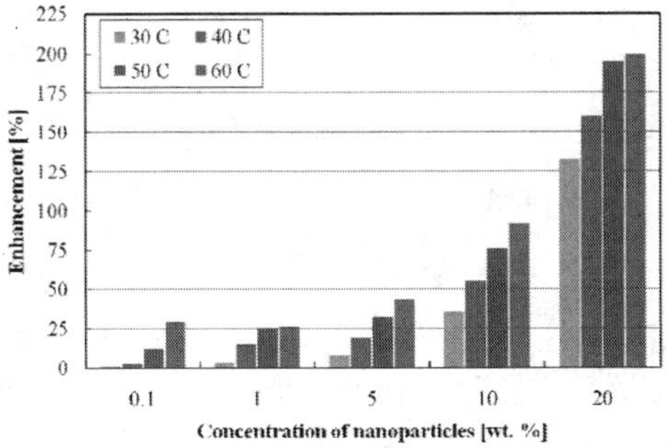

Figure 5. Enhancements in viscosity of silica nanofluids

The enhancements in the viscosity of the nanofluids were found to be larger at higher temperature. At 1% mass concentration, for instance, the viscosity of the nanofluid were increased from 3.7 % at 30 °C to 26.3 % at 60 °C. Both the concentration effect and the temperature effect are shown in Figure 5. The enhancements were more significant both for higher temperature and higher nanoparticle concentration.

DISCUSION

We observed that the viscosity of the solvent was increased by adding nanoparticles. As described above, the enhancement in the viscosity was more significant at higher temperature and particle concentrations. However, our observation about larger enhancements at higher temperature is controversial. Compared with a prior study by Namburu et al.[17], the viscosity enhancement of silica nanofluids in which mixture of water and ethylene glycol was used as a base fluid was slightly decreased with increasing temperature. Regarding this controversy between literature and our measurements, the difference in the instruments used in both studies might result in that difference trend in the viscosity enhancements. Actually, the parallel-disk type Rheometer was employed in this work and the external convection heaters played a role to maintain the temperature uniformly during the measurements. Therefore, suspension has to be exposed to surrounding so that it is possible for the base fluid to evaporate. At relatively high temperature in our temperature conditions, it is more likely for the fluid to be vaporized faster and the suspension was probably not covering the whole disk as shown in Figure 6 (B). Reduced area of the suspension between two plates could result in an elevated viscosity. Since we performed these measurements where the sweeping shear rates were varied from 100 to 1000 [1/s], the effect of area shrinkage would be more dominant at the high shear rates.

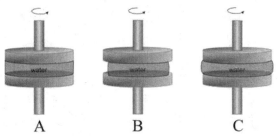

A   B   C

Figure 6. Configurations of test section: (A) Ideal case, (B) Dehydration of test material, and (C) Effect of surface tension of water.

In previous studies, it was shown that addition of nanoparticles changed the viscosity behavior of solvents to non-Newtonian (Shear-thinning fluid) [17,18]. Figure 6 (C) shows the schematic configuration of a test section which illustrates the influence of surface tension of water. Surface tension affects the fluctuations in viscosity of water and water-based nanofluids. We also tested ethylene glycol from 0.1 to 1000 to verify our facility. Ethylene glycol showed a more uniform viscosity over a wide range of shear rate. In spite of that, its viscosity was fluctuating at low shear rate less than 10. We concluded that small torque resulted in extremely high values for the measured viscosity values. Similarly, the fluctuations and the huge values in water were also observed for small values of the imposed torques. Thus, we chose the narrow shear rate from 100 to 1000 [1/s]. For this reason, we could not examine shear-thinning behavior of the nanofluids in this study. Even though the small range in shear rates was selected, inconsistent viscosity (non-Newtonian feature) was obtained from the nanofluids and its intensity was increased with nanoparticle concentrations.

CONCLUSION

In this study, viscosity of the silica nanofluids was measured using a rotational parallel-disk Rheometer. Silica (SiO_2) nanofluids samples with six different mass concentrations were prepared by diluting a nanofluid with a stock solution of 30% mass concentration. Due to some limitations in the facility used, the viscosity was measured in a narrow range of shear rate from 100 to 1000 [1/s]. In order to examine the effect of particle concentration on the rheological property, the six concentrations were tested and compared with pure water. Temperature was also varied between 30 °C and 60 °C by using external convection heaters.

The experiments showed that adding nanoparticles enhanced the viscosity of water by up to 200 % for 20% mass concentration. In relatively low mass concentrations (0.1% and 1%), the increase of the viscosity was not significant, but large enhancements were obtained from high concentration nanofluids (higher than 5%). Moreover, the enhancements were higher with increase in temperatures. To summarize, throughout this study, viscosity enhancements were observed in aqueous nanofluids. Additionally, the effects of both the concentration and temperature were also examined in this study.

ACKNOWLEDGEMENTS

For this study the authors acknowledge the support of the Department of Energy (DOE) Solar Energy Program (Golden, CO) under Grant No. DE-FG36-08GO18154; Amd. M001 (Title: "Molten Salt-Carbon Nanotube Thermal Energy Storage For Concentrating Solar Power Systems"). The authors also acknowledge the support of TA Instruments for providing the rheometer used in this study.

REFERENCES

[1]S.U.S. Choi, Developments and applications of non-Newtonian flows, *ASME FED*, **66**, 99-105 (1995)
[2]J. A. Eastman, S.U.S. Choi, and S. Li, Anomalously Increased Effective Thermal Conductivities of Ethylene Glycol-Based Nanofluids Containing Copper Nanoparticles, *Appl. Phys. Lett.*, **78**, 718-720 (2001)
[3]S.U.S. Choi, Z. G. Zhang, and W. Yu, Anomalous Thermal Conductivity Enhancement in Nanotube Suspensions, *Appl. Phys. Lett.*, **79**, 2252-4 (2001).
[4]S. K. Das, N. Putra, and P. Thiesen, Temperature Dependence of Thermal Conductivity Enhancement for Nanofluids, *J. Heat Transf.*, **125**, 567-74 (2003).
[5]S. M. S. Murshed, K. C. Leong, and C. Yang, Enhanced Thermal Conductivity of TiO_2 - Water Based Nanofluids, *Int. J. Therm. Sci.*, **44**, pp. 367-73 (2005).
[6]D. Wen and Y. Ding, Effective thermal conductivity of aqueous suspensions of carbon nanotubes (Carbon nanotube nanofluids), *J. Thermophys. Heat Transfer*, **18**, 481-5 (2004).
[7]S. M. You, J. H. Kim, and K. H. Kim, Effect of Nanoparticles on Critical Heat Flux of Water in Pool Boiling Heat Transfer, *Appl. Phys. Lett.*, **83**, 3374-6 (2003).
[8]H. D. Kim, J. Kim, and M. H. Kim, Experimental Studies on CHF Characteristics of Nano-Fluids at Pool Boiling, *Int. J. Multiphas. Flow*, **33**, 691-706 (2007).
[9]V. Sathyamuthi, H-S. Ahn, D. Banerjee, and S. C. Lau, Subcooled pool boiling experiments on horizontal heaters coated with carbon nanotubes, J. Heat Transf., **131**, 071501-10 (2009).

[10]B. Jo, P. S. Jeon, J. Yoo, and H. J. Kim, Wide Range Parametric Study for the Pool Boiling of Nano-Fluids with a Circular Plate Heater, *J. Visual. Japan*, **12**, 37-46 (2009).

[11]H. U. Kang, S. H. Kim, and J. M. Oh, Estimation of thermal conductivity of nanofluid using experimental effective particle volume, *Exp. Heat Transfer*, **19**, 181-91 (2006).

[12]C. T. Nguyen, F. Desgranges, N. Galanis, G. Roy, T. Mare, S. Boucher, and H. A. Mintsa, Viscosity data for Al2O3/water nanofluid-hysteresis: is heat transfer enhancement using nanofluid reliable?, *Int. J. Therm. Sci.*, **47**, 103-11 (2008).

[13]S. M. S. Murshed, K. C. Leong, and C. Yang, Investigations of thermal conductivity and viscosity of nanofluids, *Int. J. Therm. Sci.*, **47**, 560-8 (2008).

[14]X. Wang, X. Xu, and S.U.S. Choi, Thermal conductivity of nanoparticles-flluid mixture, J. Thermophys. Heat Transfer, 13, 474-80 (1999).

[15]H. Xie, L. Chen, and Q. Wu, Measurements of the viscosity of suspensions (nanofluids) containing nanosized Al_2O_3 particles, *High Temp.-High Press.* **37**, 127-35 (2008).

[16]R. W. Fox, A. T. McDonald, and P. J. Pritchard, Introduction to Fluid Mechanics six ed., John Wiley & Sons (2003).

[17]P. K. Namburu, D. P. Kulkarni, A. Dandekar, and D. K. Das, Experimental investigation of viscosity and specific heat of silicon dioxide nanofluids, *Micro & Nano letters*, **2**, 67-71 (2007).

[18]H. Chen, Y. Ding, A. Lapkin, and X. Fan, Rheological behavior of ethylene glycol-titanate nanotube nanofluids, *J. Nanopart. Res.*, **11**, 1513-20 (2009).

PUMPING POWER OF 50/50 MIXTURES OF ETHYLENE GLYCOL/WATER CONTAINING SiC NANOPARTICLES

Jules L. Routbort[1], Dileep Singh[2], Elena V. Timofeeva[1], Wenhua Yu[1], David M. France[3], and Roger K. Smith[1]

[1]Energy Systems Division, Argonne National Laboratory, Argonne, IL 60439, USA
Tel: 1-630-252-5065, Fax: 1-630-252-5568, Email: routbort@anl.gov

[2]Nuclear Engineering Division, Argonne National Laboratory, Argonne, IL 60439, USA

[3]Deptment of Mechanical & Industrial Engineering, University of Illinois at Chicago, Chicago, IL 60607, USA

ABSTRACT

The torque (proportional to the pumping power) of nanofluids containing 2.2 vol. % 29-nm and 4.0 vol.% 90-nm SiC nanoparticles in a 50/50 mixture of ethylene glycol/water have been measured in the turbulent flow regime at 30±1°C. The results confirm the theoretical calculations of pumping power suggesting that the nanofluid can be treated as a single-phase material in a pumping system with elbows, expanders, and straight pipes. They also confirm the experimental finding that the torque for larger particles is smaller than for smaller particles, which is in agreement with the experimental result on the nanofluid viscosity decreasing with increase in the particle sizes.

INTRODUCTION

One of the obstacles to practical application of nanofluids is that addition of nanoparticles to a fluid increases its viscosity[1]. The increased viscosity results in an increase in the power required to pump the nanofluid at the same velocity as the base fluid, as has been reported for 5 vol. % of Al_2O_3, CuO, and diamond nanoparticles in water[2]. Therefore, the increase in heat removal, resulting from the enhanced thermal conductivity and heat transfer coefficient of nanofluids that could result in a decrease in the size of the heat exchanger and the accompanying increase in fuel efficiency for a vehicle, might be offset by the increase in power required to pump the nanofluid. We have, therefore, built a fluid pumping system to measure the torque required to pump nanofluids. The experimental results are compared to calculations based on the friction factors in the system.

EXPERIMENTAL

The SiC nanoparticles dispersed in a 50/50 ethylene glycol/water (EG/W) mixture have been well characterized[3] with respect to volume concentration and particle sizes (16–90 nm). Additionally, the viscosity, thermal conductivity, and heat transfer coefficient have been measured and correlated with concentration and particle size. The conclusion of that work was that, at the same volume concentration, the viscosity is lower in suspensions with larger particles, and the viscosity increases due to the addition of nanoparticles while it decreases with temperature. Thus the viscosity of the 4 vol. % 90-nm SiC nanofluid is nearly equal to that of the base fluid at about 75°C while the

147

viscosity of the 66-nm nanoparticles equals that of the base fluid at about 100°C. The heat transfer enhancement of the 4 vol. % 90-nm SiC is ~15% whereas the smaller particles result in a lower enhancement[3].

A schematic of the experimental apparatus is shown in Fig. 1. It is typical of vehicle pump systems containing heat exchangers. The torque meter (not shown in the figure) is mounted on the pump shaft. For measuring torque, valves 3, 5, and 7 are shut. The non-contact torque meter is manufactured by Advanced Telemetrics (Spring Valley, Ohio) using a wireless transmitter. The flow rate is measured by an electromagnetic flow meter (Proline Promag 10H) manufactured by Endress+Hauser (Greenwood, Indiana). Both the flow and the torque meters were calibrated, the former by measuring the fluid volume flow for a predetermined time, the latter by the manufacturer. Flow velocities are in the range of 20–28 l/min (1400–1800 RPM) corresponding to Reynolds numbers (Re) of ≈7000, representing the turbulent flow regime.

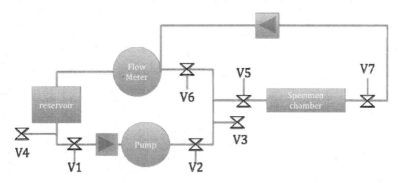

Figure 1. Schematic of the torque apparatus. Torque meter and data logger are not shown. A photograph of the actual apparatus appears in ref. 4.

Torque measurements were made continuously over a period between 10 and 40 minutes at each flow velocity and were the average values of hundreds of measurements with each data point representing 4 seconds of accumulation. The data were collected by a PC using LabView software. The deviation of the torque at a given flow rate for the SiC nanofluids or the pure EG/W mixture was less than ± 3%. The pumping speed (RPM) of the pump was measured optically.

The torque τ and the angular velocity ω, measured directly, are used to calculate the shaft power P_{shaft}, which can be expressed as a function of the pumping power $P_{pumping}$ through the pump efficiency η

$$P_{shaft} = \tau\omega = P_{pumping}/\eta \qquad (1)$$

In the above equation, the pumping power for a fluid with density ρ and viscosity μ can be expressed as the product of the volumetric flow rate Q and the pressure drop Δp

$$P_{pumping} = Q\Delta p \qquad (2)$$

where the volumetric flow rate can be expressed as a function of velocity V through the cross section area A or the circular tube diameter d

$$Q = AV = (\pi d^2/4)V \qquad (3)$$

For a continuously flowing system, the pressure drop usually includes the static head as a function of vertical position h ($\Delta p_{static} = \rho g h$), the friction pressure drop as a function of the friction factor f and the tube length L ($\Delta p = 2\rho V^2(L/d)f$), and other pressure losses due to elbows, expansions, contractions, and a variety of other components. By expressing the component pressure losses in terms of the component inlet velocity V_i, the shaft power for a turbulent fluid flow[5] can be expressed as

$$P_{shaft} = \frac{Q\Delta P}{\eta} = \frac{(\pi d^2/4)V[\rho g h + 2\rho V^2(L/d)f + \sum_i K_i(1/2)\rho V_i^2]}{\eta} \qquad (4)$$

where the pressure loss coefficient K_i is determined by the nature of a pressure loss component and the Fanning friction factor f depends on the flow conditions and can be estimated through the Reynolds number Re by the Blasius equation[6] for a turbulent fluid flow as

$$f = 0.0791\,\text{Re}^{-0.25} = 0.0791(\rho V d/\mu)^{-0.25} \qquad (5)$$

A simplified flow model was developed for the system shown schematically in Fig. 1 consisting of a straight smooth pipe (2 m in length), five small elbows with 90° radius, and two abrupt expansions. These eight components were used as the model for calculating the pressure drop through the system per Eq. 4. The pumping power of a nanofluid was calculated from the the pressure drop and the volumetric flow rate, and the the ratio was taken of the pumping power of the nanofluid to the base fluid at the constant velocity for all cases. That ratio was rather insensitive to the simplicity of the flow model of the experimental system. For example, doubling the pipe length in the model (a marked deviation from reality) produced almost no measurable changes in the pumping power ratio prediction. Therefore, the simplified model adequately describes the actual pumping system.

RESULTS

The torque measured for the 4 vol. % 90-nm SiC nanoparticles in a 50/50 mixture of EG/W is shown in Fig. 2 where the data are compared to the pure 50/50 mixture of EG/W. The two linear regression lines are approximately parallel over the entire range of velocities used. Therefore we arbitrarily choose $V = 25$ l/min to compare the relative torques τ_{nf}/τ_{bf} where nf represents the fluid containing the nanoparticles and bf

represents the base fluid, 50/50 mixture of EG/W. The results for the two SiC nanofluids are given in tabular form in Table 1 below. The comparison of the experimental and calculated values shows good agreement well within the uncertainty of the data.

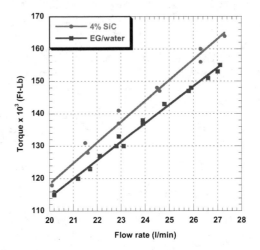

Figure 2. Measured torque of the fluid containing 4 vol. % 90-nm SiC nanoparticles compared to the pure base fluid at the constant velocity.

Table 1. Comparisons of the calculated and experimental relative percent increase in torque.

Volume % SiC (particle size)	Relative % increase in torque – Calculated	Relative % increase in torque – Experimental
2.2 (29 nm)	8.7±0.3	9.3±0.3
4.0 (90 nm)	6.5±0.3	5.3±0.3

DISCUSSION

There are two important observations. The first is that the calculated relative percent increase in torque and the experimental values *agree within the uncertainties*. This is consistent with the previous results published for water-based nanofluids of Torii[2] and Williams and Buongiorno[7] using an Al_2O_3/water-based nanofluid[4]. However, the previously cited work was performed on water-based systems using straight tubes only. We have preformed these experiments using a system that more closely approximates a complete vehicle system using a higher viscosity-base fluid, 50/50 mixture of EG/W. An important result is that one can calculate the torque (pumping power) in the turbulent region by considering the nanofluid as a single phase fluid.

The second observation is that the torque of the nanofluid containing the larger nanoparticles is lower than that of the nanofluid containing smaller particles. This was to be expected as the viscosity of the fluid containing the 90-nm SiC is lower at all temperatures than that of the 29-nm SiC nanofluid[3]. This observed behavior is contrary to the classical Einstein-Bachelor equation for hard non-interacting spheres that predicts that the viscosity is only a function of the volume particle concentration. The fact that the viscosity is dependent on particle size has been attributed to the difference in structure and thickness of the diffuse fluid layers around the nanoparticles in various base fluids that affects the effective volume concentration[3].

Finally, it should be noted that fluids containing larger nanoparticles have higher heat transfer rates than those containing smaller nanoparticles as a consequence of lower surface area for the same volume concentration of nanoparticles[3]. Smaller area of solid/liquid interface in nanofluids results in a smaller negative contribution of the interfacial thermal resistance and higher thermal conductivity of the fluids with larger nanoparticles.

It is also worth noting that because the viscosity of the 66- and 90-nm nanoparticles becomes equal to that of the base fluid at 100°C and 75°C, respectively, the pumping power losses for nanofluids will be eliminated at the operating temperatures of many vehicle heat exchangers.

CONCLUSIONS

Torque required to pump a nanofluid in the turbulent flow region in vehicular systems can be calculated from the pressure drop and the volumetric flow rate (assuming constant pump efficiency) by considering the nanofluid as a single phase fluid. The increase in pumping power for the nanofluids with larger SiC nanoparticles in EG/W at heat exchanger operating temperatures will be insignificant compared to the EG/W alone.

ACKNOWLEDGEMENTS

The authors are grateful to Dr. Steve Hartline of Saint-Gobain for supplying the SiC-water nanofluid. This work was sponsored by the Office of Vehicle Technologies and the Industrial Technology Program of the U.S. Department of Energy under contract number DE-AC02-06CH11357.

REFERENCES

1. R. D. Vold, M. J. Vold, Colloid and interface chemistry, Addison-Wesley, Reading, MA, 1983.
2. S. Torii, Turbulent heat transfer behavior of nanofluid in a circular tube heated under constant heat flux, Advances in Mechanical Engineering, Article 1D 917612, 2010.
3. E. V. Timofeeva, W. Yu, D. M. France, D. Singh, J. L. Routbort, Base fluid and temperature effects on the heat-transfer characteristics of SiC nanofluids in EG/H$_2$O and H$_2$O, Journal of Applied Physics, **108**, 2010, in press.
4. J. L. Routbort, D. Singh, E. Timofeeva, W. Yu, D. M. France, Pumping power of nanofluids in a flow system, Journal of Nanoparticle Research, 2011, in press.
5. R. Fox, A. McDonald, P. Pritchard, Introduction to fluid mechanics, ed. 7, John Wiley & Sons, NJ, 2008.

6. H. Blasius, Das Ähnlichkeitsgesetz bei Reibungsvorgängen in Flüssigkeiten, Forschungsarbeiten des Ingenieurwesens, Heft 131, Verein Deutscher Ingenieure, Berlin, 1913.

7. W. Williams, J. Buongiorno, L.-W. Hu, Experimental investigation of turbulent convective heat transfer and pressure loss of alumina/water and zironia/water nanoparticle colloids (nanofluids) in horizontal tubes, Journal of Heat Transfer, 130, paper 04241, 2008.

Computational Design

CHARACTERIZATION OF NON UNIFORM VENEER LAYER THICKNESS DISTRIBUTION ON
CURVED SUBSTRATE ZIRCONIA CERAMICS USING X-RAY MICRO-TOMOGRAPHY

M. Allahkarami[1], H. A. Bale[2], J. C. Hanan[1*]

[1]Mechanical and Aerospace Engineering, Oklahoma State University

Tulsa, OK, USA

[2] Material Science and Engineering, University of California

Berkeley, CA, USA

ABSTRACT

The fabrication process of bi-layer ceramics requires a temperature change from sintering to room temperature. Residual stresses created as a result of thermal expansion coefficient mismatch may be sufficiently large to influence failure. Theoretical models predict residual stress as a function layer thickness. The scale and to some extent the geometry changes during the firing process. In the case of complex geometries like dental crowns, micro-X-ray tomography is a unique non-destructive technique for layer thickness measurement. A method of developing a finite mesh model of real components involving complex geometries has been developed through the use of X-ray micro-tomography. This method computes the minimum distance between two triangulated surfaces constructed from tomography data. For each vertex of one surface, it computes the closest point on the other surface. The veneer layer thickness distribution was combined with residual stress curves advancing our understanding of the lower limit threshold of thickness in dental restoration design. An important observation is the significant range in predicted residual stress in the veneer top surface from a tensile stress of 80 MPa to a compressive stress of -70 MPa as the thickness increased to 0.5 mm.

INTRODUCTION

A wide range of industrially important materials, from aircraft engine parts to dental crown ceramics, are made of bi-layer or multi layer structures. The application of layered ceramics or metal ceramic components is challenged by thermal expansion coefficient mismatch that creates curvature effects and residual stress [1,2]. Timoshenko fist derived a general solution for bi-layer bending due to residual stresses [3]. Considerable efforts [4,5,6] have been devoted to analyzing residual stresses for multilayer systems because of their wide application as microelectronic, optical, and structural components. Residual stress in the interface of a bi-layered material is a function of layer thickness [7]. In the case of a dental crown, the bi-layer is not thin compared with the curvature radius of the surface, but the state of residual stress can be estimated from the local thickness of zriconia and porcelain layers.

Here we focus on ceramic dental crowns. The base layer almost has a constant thickness, typically around 0.5 mm, and the thickness of porcelain is varied to create the aesthetic and functional geometry of the crown. Micro X-ray tomography slides have been used to reconstruct the geometry of crowns for finite element analysis [8,9,10] Here in this paper we will explain a method that utilizes tomography data to reconstruct the 3D structure of the crown followed by finite element surface meshing to calculate the minimum distance between nodes on surfaces. Thickness measured using this

---

* Corresponding author; e-mail: jay.hanan@okstate.edu; phone: 918-594-8238.

method can be coupled by bi-material bending equations to approximate the state of residual stress at different locations of the crown.

METHOD

Micro-Computed Tomography data, collected by H. A Bale and J. C Hanan [10], was used for 3D reconstruction of a bi layered zirconia porcelain. The following discusses a method used to calculate the local thickness of the porcelain layer. There are two surface functions, for zirconia $f_z(x,y)$ and for porcelain $f_p(x,y)$, obtainable from X-ray tomography, as illustrated in Figure 1. Because $f_z(x,y)$ and $f_p(x,y)$ are two locally variant curved functions, the distance between them is not well defied. It is possible to calculate the shortest surface-to-surface distance for the grid points laying on one of the surfaces. Here we are interested to measure the shortest distance between each point on $f_z(x,y)$, the base surface, relative to the outside of the porcelain layer, $f_p(x,y)$ surface. Therefore, the zirconia upper surface is considered a reference surface, and the minimum distance to the porcelain surface at each point is desired. If coordinates are selected as shown in Figure 1, we are dealing with two real valued function $f_z(x,y)$ and $f_p(x,y)$ that satisfy $f_z(x,y) < f_p(x,y)$ for all point on the domain of $f_z(x,y)$. This is a z-simple geometry because the region is described in a simple way, using z as a function of x and y. This means, for each pair of "x, y" there is one unique corresponding "z" value.

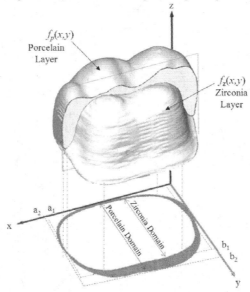

Figure 1 Graphical rendering of tomography from a bi-layer dental crown made of ceramic base covered by porcelain.

Three dimensional data visualization and finite element triangular surface meshing was carried out in Amira 3.1.1. A customized Matlab code was developed to extract the nodes that belong to $f_z(x,y)$ and $f_p(x,y)$. For a better visualization, nodes for the upper surface of the porcelain layer are colored green in Figure 2, and nodes for the interface surface of zirconia and porcelain are determined with a red color (this appears darker in black and white).

The minimum distance for a point i on the interface surface that has a coordinate of $X_i=(x_i,y_i,(f_z(x_i, y_i))$ on the domain $[a_{z1},b_{z1}]\le [x_i,y_i]\le [a_{z2},b_{z2}]$ is defined as,

$$d(x_i,y_i,z_i) = \min\{\sqrt{[(x_i-x_j)^2+(y_i-y_j)^2+(f_z(x_i,y_i)-f_p(x_j,y_j))^2]}\}\ ;\ \text{for all possible j}\} \quad (1)$$

Where j is the index for a point j on the interface surface (zirconia surface) that has a coordinate of $X_j=(x_j,y_j,(f_p(x_j, y_j))$ on the domain of $[0,0]\le [x_j,y_j]\le [a_{p2},b_{p2}]$. The distance $d(x_i,y_i,z_i)$ is always positive real number and is equal to zero at intersecting points or overlapped regions. A histogram of the distance measurement is illustrated in Figure 3.

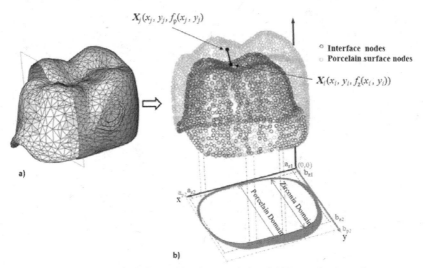

Figure 2 a) Triangular surface meshing of tomographed crown b) Extracted surface nodes.

Because the density of nodes in a certain area is a function of surface fluctuation, nodes are randomly distributed on the interface and veneered surface. The region with more surface fluctuation has more density of nodes in comparison with flat regions. This has not affected the accuracy of the minimum distance measurement for a particular node but introduces an error in the histogram of distance distribution, because we are not sampling the domain with equal probability.

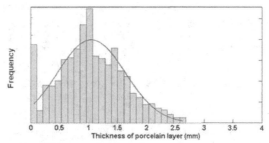

Figure 3 Histogram of minimum distance measurement

Corrections on Minimum Distance Calculation Using 3D Interpolation

Increasing the number of nodes relatively decreases this error and increases computing time. Although computation time is reducible by applying a constraint on the searching area to a ball with radius of r_0 instead of global searching for minimum distance, increasing the number of nodes does not guarantee equal probability of node distribution; and sharp edges and regions faced toward z direction will have more contribution.

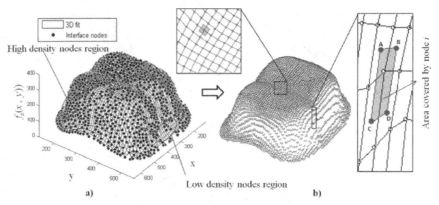

Figure 4 a) Randomly distributed nodes (blue markers) with a 3D interpolated surface (black lines). b) Generated regular nodes with an illustration of the area covered by nodes in a relatively flat and a sloped region.

A simple solution to this is applying 3D interpolation between the nodes to create a regular mesh mapped on the surface. Using the same method, new sets of nodes can be created for the porcelain surface as well. The result of 3D interpolation for interface surface and a new set of nodes, selected as the edge of a grid that covers up the surface, are illustrated Figure 4a and b, respectively. These two new sets of nodes are distributed with equal density on their projection to the x-y plane. In 3D space the area covered by each node is depend on the slope of the tangent plane to the node. In result, nodes located on the walls of the crown will cover more surface area in space. If we multiply the minimum distance calculated for each node by a unitless weight, corresponding to the area that the node covers, the resulting histogram of the weighted distance distribution is corrected from the prior node distribution error. In fact, decomposing the surface between the nodes is an implementation of Voronoi tessellation decomposition of the surface with the seeds located at the nodes. The surface area corresponding to a node located in a point $X_i=(x_i,y_i,(f_z(x_i, y_i))$ can be approximated by the average of four cross product absolute values, if we have the coordinate of the tetragon vortex that contains the node. These required coordinates are labeled with A, B, C and D in Figure 4b and are computed using 3D interpolation.

$$S(x_i,y_i,f_z(x_i,y_i)) = \frac{1}{4}\left(\left|\overrightarrow{AB}\times\overrightarrow{AC}\right|+\left|\overrightarrow{AB}\times\overrightarrow{DB}\right|+\left|\overrightarrow{BD}\times\overrightarrow{DC}\right|+\left|\overrightarrow{AC}\times\overrightarrow{CD}\right|\right) \qquad (2)$$

The shortest surface-to-surface distance along the interface surfaces, and the area belonging to the node $S(x_i,y_i,(f_z(x_i, y_i))$, were calculated for each node using a Matlab code.
The profile of a calculated Voronoi cell area for nodes along the cross section is illustrated in Figure 5 There are two peaks at both sides of the profile corresponding to the wall area of crown which has more slope and less density of nodes. The flat region between two sharp peaks is related to the top surface of the crown which has greater areal density of nodes.

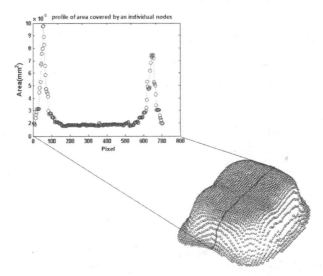

Figure 5 Profile shows the area corresponding to the nodes.

In order to better visualize the effect of the nodes distribution density on thickness calculations, a contour plot of the area belonging to the individual nodes is shown in Figure 6a. Figure 6b is the histogram corresponding to the contour plot. It can be seen, most of the areas belonging to the top portion of the crown have the same range, but values are much higher for the walls, which confirm the importance of the node distribution correction.

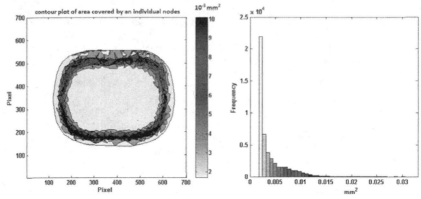

Figure 6 Contour plot (top view) and histogram of area covered by nodes

For weighted by area nodes, we cannot simply plot the histogram, but it is possible to calculate the Probability Density Function (PDF), which here is a function that describes the relative likelihood for the thickness of a layer, as a variable, to occur at a given point in the observation space. The PDF of minimum distance weighted by node area is shown in Figure 7. This PDF thickness distribution is independent of node density distribution.

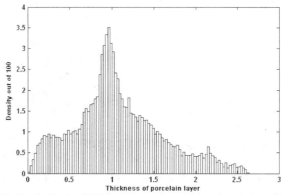

Figure 7 Probability Density Function (PDF) for thickness of porcelain layer in typical porcelain covered zironia base dental crown.

Using the model previously described by the authors [7], it is possible to predict the effect of the porcelain layer thickness (here for a fixed zirconia layer at 0.5 mm) on residual stress in particular positions of interest for the bottom of the zirconia layer, interface of the zirconia side, interface of the porcelain side, and top surface of the porcelain; as shown in Figure 8a.

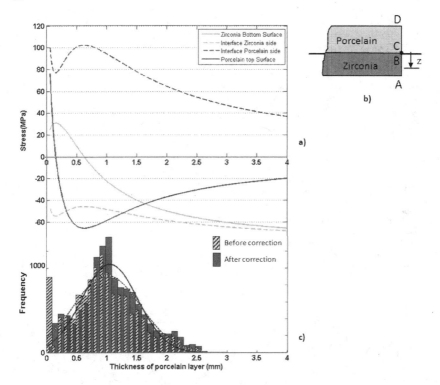

Figure 8 a) Residual stresses in the interface and top and bottom surface as a function of porcelain thickness. b) The locations A, B, C, and D with respect to the bi-layer geometry and 'z' indicates the thickness of zirconia and porcelain. c) Histogram of minimum distance measurement before and after correction

Histogram of minimum distance measurement (Figure 8c) was corrected using PDF information. Comparing histogram of thickness distribution before and after this correction reveals a considerable change at thin thickness region (less than 0.2 mm). Error in distribution introduced mostly in thin regions, because locations with thin porcelain layer thickness have more complexity in geometry variation. In general meshing software dedicates more number of nodes to maintain the geometry variations in complex geometry regions which later caused the error in thickness distribution. Using described technique correct thickness distribution was measured.

CONCLUSIONS

A method of generating surface meshing for finite elements using micro X- tomography of dental crowns was utilized to extract node coordinates on the surface of real crown geometry. The result of this method was used to calculate the thickness of the porcelain layer in a zirconia-porcelain bi-layer ceramic crown. A histogram of the thickness distribution was combined with a beam bending model residual stress prediction for bi-layer ceramics keeping the thickness of the zirconia layer constant. Combining the model with the thickness measurement suggests development of crowns with controlled porcelain thickness in order to minimize residual stress. Over all, locations where the thickness of the porcelain relative to the zirconia layer is small (<0.2 mm) should be avoided. This is generally the case for the observed crown as only 3.3% of the porcelain was below this thickness. 53% was above 1 mm, which appears a safer thickness when considering the predicted residual stresses. It was found that for this typical molar crown, the thickness of the porcelain layer changed from 0.05 mm to 2.5 mm, and the most probable 43% thickness was around 1±0.2 mm.

ACKNOWLEDGEMENTS

Partial research support provided by the Oklahoma Health Research award (project number HR07-134), from the Oklahoma Center for the Advancement of Science and Technology (OCAST) and the Helmerich Advanced Technology Research Center, Oklahoma State University, Tulsa.

REFERENCES

[1] Jurgen Malzbender, Curvature and stresses for bi-layer functional ceramic materials, *Journal of the European Ceramic Society*, **30**, 16, 3407-3413, (2010)

[2] Addison, O. and G. J. P. Fleming, Application of analytical stress solutions to bi-axially loaded dental ceramic-dental cement bilayers, *Dental Materials*, **24** (10), 1336-1342, (2008).

[3] S. Timoshenko, Analysis of Bi-Metal Thermostats, *J. Opt. Soc. Amm.*, **11**, 233–255 (1925).

[4] C. H. Hsueh, S. Lee, T. J. Chuang, An Alternative Method of Solving Multilayer Bending Problems, *Journal of Applied Mechanics.*, **70** 151-154 (2003).

[5] C. H. Hsueh, C. R. Luttrell, P. F. Becher, Analyses of multilayered dental ceramics subjected to biaxial flexure tests, *Dental Materials.*, **22**, 460–469 (2006).

[6] C. H. Hsueh, G. A. Thompson, O. M. Jadaan, A. A. Wereszczak, P. F. Becher, Analyses of layer-thickness effects in bilayered dental ceramics subjected to thermal stresses and ring-on-ring tests, *dental materials.*, **24**, 9–17 (2008).

[7] M. Allahkarami, H. A. Bale, J. C. Hanan, Analytical model for prediction of residual stress in zirconia porcelain bi-layer, *34th international conference on advanced ceramics and composites (ICACC) Daytona*, (2010)

[8] Verdonschot, N., Willem M. M. Fennis, Ruud H. Kuijs, Jan Stolk, Cees M. Kreulen, Nico H. J. Creugers, Generation of 3-D finite element models of restored human teeth using micro-CT techniques, *International Journal of Prosthodontics*, **14**(4), 310-315 (2001)

[9] Magne, P., Efficient 3D finite element analysis of dental restorative procedures using micro-CT data, *Dental Materials*, **23**, 539-548, (2006)

[10]H. A. Bale, P. G. Coelho, J. C. Hanan, 3D Finite Element Prediction of Thermal Residual Stresses in a Real Dental Restoration Geometry, *In Progress* (2010)

COMPUTATIONAL STUDY OF WAVE PROPAGATION IN SECOND-ORDER NONLINEAR PIEZO-ELECTRIC MEDIA

David A. Hopkins, George A. Gazonas
U.S. Army Research Laboratory
Weapons and Materials Research Directorate
Aberdeen Proving Ground, MD 21005

ABSTRACT

In this paper, we present computational results for the response of piezoelectric materials with 6mm symmetry when subjected to shock loads. The governing equations are based on a second-order theory of piezoelectricity for the 6mm crystal class formulated in a Lagrangian reference configuration. These equations represent the fully coupled nonlinear multiphysics response. Numerical solutions of these equations are first verified using analytical solutions for wave propagation in linear piezoelectric media. The effects of the nonlinear coupling introduced by higher order elastic, dielectric and piezoelectric coupling coefficients are then examined. The wave speed is shown to be lower than the bar velocity predicted by linear piezoelectricity for large strains.

INTRODUCTION

The direct piezoelectricity effect in which a mechanical load generates an electric charge was first discovered by the Curie brothers in 1880. The inverse effect where an electric field generates a mechanical response was predicted by Lippmann in 1881 and subsequently confirmed by the the Curies[1]. Subsequently, the behavior of piezoelectric materials in non-structural applications has been investigated extensively though these investigations almost exclusively employ linear constitutive relations. Piezoelectricity is currently enjoying a great resurgence in both fundamental research and technical applications[2-6]. Investigations in the nonlinear basic theory for piezoelectricity though have been limited. Nelson[7], Toupin[8] and Tiersten[9] studied the nonlinear theory of dielectrics. Norwood et al.[10], Kulkarni and Hanagud[11] used a Neo-Hookean constitutive relation to model the response of piezoelectric ceramics. Pai et al.[12] considered the dependence of the piezoelectric strain parameters upon the strain in formulating a plate theory of piezoelectric laminates. Joshi[13] considered the nonlinear constitutive relations for piezoelectric materials, where a concise expression was given. Tiersten[14] investigated the nonlinear problems of thin plates subjected to large driving voltages. However, most of these studies still only considered the case of small deformations, and second-order effects were often neglected. Later on, based on the theory of invariants, from invariant polynomial constitutive relations, Yang and Batra[15] investigated the second-order theory for piezoelectric materials with symmetry class 6mm and class mm2. Feng, et al.[16], using results from Kiral and Eringen[17], developed the relations for symmetry classes 6mm and 3m. In this paper, we present these equations in a form suitable for numerical solution. These equations are then specialized for 1D. The resulting equations are first verified by comparison with solutions obtained using numerical inverse Laplace transform techniques. Finally, the response of the full nonlinear equations to step pressure representative of a shock is shown and discussed.

THEORY

For a continuous body, the 3D equation of motion in the current (Eulerian) configuration is given by Malvern[18]

$$\frac{\partial T_{kl}}{\partial x_k} + \rho b_l = \rho \ddot{u}_l \quad , \tag{1}$$

where ρ is the material density, b_l is the body force, $\ddot{u}_l$ is the particle acceleration, and T_{kl} is the total stress. Lower case Latin subscripts indicate quantities are referred to the current (Eulerian) configuration while upper case Latin subscripts indicate quantities are referred to the reference (Lagrangian) configuration with subscripts ranging from $1-3$. The total stress, T_{kl}, can be written[15]

$$T_{kl} = T_{kl}^C + \varepsilon_0 \hat{E}_k \hat{E}_l - \frac{1}{2}\varepsilon_0 \hat{E}_m \hat{E}_m \delta_{kl} \quad . \tag{2}$$

T^C is the Cauchy stress and the remaining terms are the Maxwell stress expressed in the current (Eulerian) configuration[19] where ε_0 is the electric permittivity of free space, $\hat{E}$ is the electric field, and δ_{kl} is the Kronecker delta. The equations of motion in the reference (Lagrangian) configuration are obtained from (1) with appropriate transformations. First, the equivalence of the divergence in different reference frames is given by

$$(G_{kl})_{,k} = \frac{1}{J}(JF_{Kk}^{-1}G_{kl})_{,K} \quad , \tag{3}$$

where G is an arbitrary second rank tensor. The Cauchy stress, T_{kl}^C, is expressed in terms of the second Piola-Kirchoff stress, T_{KL} by

$$T_{kl}^C = \frac{1}{J}x_{k,K}T_{KL}x_{l,L} \quad , \tag{4}$$

and the electric field and polarization transform via[20]

$$\hat{E}_k = X_{K,k}E_K = F_{Kk}^{-1}E_K \tag{5}$$

$$P_k = \frac{1}{J}x_{k,K}\Pi_K = \frac{1}{J}F_{kK}\Pi_K \tag{6}$$

where E is the electric field in the reference configuration and P and Π are the polarization in the current and reference configurations, respectively, and $F_{kK} \equiv x_{k,K}$ is the deformation gradient. It is worth noting that researchers have used different transformations for P_k[21,22]. Using the relations $F_{Kk}^{-1} \equiv X_{K,k}, J\rho = \rho_0$ where ρ_0 is the density referred to the reference configuration with [3-6], the equation of motion is

$$(T_{KL}x_{l,L})_{,K} + (JF_{Kk}^{-1}(\varepsilon_0 F_{Mk}^{-1}F_{Nl}^{-1} - \frac{1}{2}\varepsilon_0 F_{Mm}^{-1}F_{Nm}^{-1}\delta_{kl})E_M E_N)_{,K} + \rho_0 b_l = \rho_0 \ddot{u}_l \quad . \tag{7}$$

Similarly, starting with Gauss's law,

$$(D_k)_{,k} = 0 \quad , \tag{8}$$

where D is the electric displacement, and using the constitutive relationship $D_k = \varepsilon_0 \hat{E}_k + P_k$, (8) becomes

$$(\varepsilon_0 JF_{Kk}^{-1}F_{Mk}^{-1}E_M + \Pi_K)_{,K} = 0 \quad . \tag{9}$$

The Euler-Piola-Jacobi identity, $(JF_{Kk}^{-1})_{,K} = 0$, is used to simplify (9) to

$$(\Pi_K)_{,K} + \varepsilon_0 JF_{Kk}^{-1}(F_{Mk}^{-1}E_M)_{,K} = 0 \quad . \tag{10}$$

Equations (7) and (10) are the equation of motion and Gauss's law in the reference configuration. Closed form solutions to (7) and (10) are not readily obtained. Consequently, solutions are sought using the weak form. This is obtained by premultiplying the governing equations by suitable test functions and adding. Accordingly, (7) and (10) can be expressed in the weak form as[23]

$$-\int_{\Omega_0}(T_{KL}x_{l,L} + (JF_{Kk}^{-1}(\varepsilon_0 F_{Mk}^{-1}F_{Nl}^{-1} - \frac{1}{2}\varepsilon_0 F_{Mm}^{-1}F_{Nm}^{-1}\delta_{kl})E_M E_N))v_{l,K}d\Omega_0 + \int_{\Omega_0}\rho_0 b_l v_l d\Omega_0$$

$$-\int_{\Omega_0}\rho_0 \ddot{u}_l v_l d\Omega_0 + \int_{\Gamma_0}(T_{KL}x_{l,L} + (JF_{Kk}^{-1}(\varepsilon_0 F_{Mk}^{-1}F_{Nl}^{-1} - \frac{1}{2}\varepsilon_0 F_{Mm}^{-1}F_{Nm}^{-1}\delta_{kl})E_M E_N))n_K v_l d\Gamma_0 \tag{11}$$

$$-\int_{\Omega_0}(\Pi_K + \varepsilon_0 JF_{Kk}^{-1}(F_{Mk}^{-1}E_M))\Phi_{,K}d\Omega_0 + \int_{\Gamma_0}(\Pi_K + \varepsilon_0 JF_{Kk}^{-1}(F_{Mk}^{-1}E_M))n_K\Phi d\Gamma_0 = 0$$

where v_l and Φ are an arbitrary displacement and scalar electric potential test functions, respectively, Ω_0 is the domain of the reference configuration with boundary Γ_0, and n_K is the outward normal on the boundary. Assuming the response is magnetostatic, the electric field, E, can be expressed as the gradient of Φ

$$E = -\nabla\Phi \quad .$$

The variational statement can therefore be written

$$-\int_{\Omega_0} (T_{KL}x_{l,L} + (JF_{Kk}^{-1}(\varepsilon_0 F_{Mk}^{-1} F_{Nl}^{-1} - \frac{1}{2}\varepsilon_0 F_{Mm}^{-1} F_{Nm}^{-1}\delta_{kl})\Phi_{,M}\Phi_{,N}))v_{l,K}d\Omega_0 + \int_{\Omega_0} \rho_0 b_l v_l d\Omega_0$$

$$-\int_{\Omega_0} \rho_0 \ddot{u}_l v_l d\Omega_0 + \int_{\Gamma_0} (T_{KL}x_{l,L} + (JF_{Kk}^{-1}(\varepsilon_0 F_{Mk}^{-1} F_{Nl}^{-1} - \frac{1}{2}\varepsilon_0 F_{Mm}^{-1} F_{Nm}^{-1}\delta_{kl})\Phi_{,M}\Phi_{,N}))n_K v_l d\Gamma_0 \qquad (12)$$

$$-\int_{\Omega_0} (\Pi_K - \varepsilon_0 J F_{Kk}^{-1}(F_{Mk}^{-1}\Phi_{,M}))\Phi_{,K}d\Omega_0 + \int_{\Gamma_0} (\Pi_K - \varepsilon_0 J F_{Kk}^{-1}(F_{Mk}^{-1}\Phi_{,M}))n_K\Phi d\Gamma_0 = 0 \quad .$$

For small electric fields, (12) reduces to

$$-\int_{\Omega_0} T_{KL}x_{l,L}v_{l,K}d\Omega_0 + \int_{\Omega_0} \rho_0 b_l v_l d\Omega_0 - \int_{\Omega_0} \rho_0 \ddot{u}_l v_l d\Omega_0 - \int_{\Omega_0} \Pi_K \Phi_{,K}d\Omega_0$$

$$+ \int_{\Gamma_0} T_{KL}x_{l,L}n_K v_l d\Gamma_0 + \int_{\Gamma_0} \Pi_K n_K \Phi d\Gamma_0 = 0 \quad . \qquad (13)$$

T_{KL} and Π_K are related to the strains and the electric field by appropriate constitutive relations. These relations are obtained from

$$T_{KL} = \frac{\partial\Sigma}{\partial\Gamma_{KL}} \qquad (14)$$

$$\Pi_K = \frac{\partial\Sigma}{\partial E_K} \quad . \qquad (15)$$

where Σ is the free energy density function. Expressions for Σ are lengthy. Feng et al.[24], and Kiral and Eringen[17] provide explicit expressions. In 1D, these expressions reduce to

$$T = a\Gamma - eE + \frac{1}{2}C\Gamma^2 - \frac{1}{2}QE^2 - g\Gamma E \quad , \qquad (16)$$

$$\Pi = e\Gamma + \varepsilon E + \frac{1}{2}g\Gamma^2 + \frac{1}{2}\eta E^2 + Q\Gamma E \qquad (17)$$

where a is the second order elastic stiffness coefficient, e is the piezoelectric coupling coefficient, ε is the dielectric permittivity, C is the third order elastic coefficient, Q is electrostrictive coefficient, g is the third order piezoelectric coupling coefficient, and η is the third order dielectric permittivity. Γ is the finite strain measure. It is given by

$$\Gamma_{IJ} = \frac{1}{2}\left(\frac{\partial u_I}{\partial X_J} + \frac{\partial u_J}{\partial X_I} + \frac{\partial u_K}{\partial X_I}\frac{\partial u_K}{\partial X_J}\right) \quad , \qquad (18)$$

which in 1D reduces to

$$\Gamma = \frac{\partial u}{\partial Z} + \frac{1}{2}\left(\frac{\partial u}{\partial Z}\right)^2 \quad . \qquad (19)$$

Table I: Material Properties for Quartz and PZT

Material	Quartz[28]			PZT[25]
Name	Present	D&G	Value	
Elastic stiffness (GPa)	a	C_{33}^E	8.6736×10^{10}	11.541×10^{10}
Piezoelectric coupling (C/m^2)	e	e_{33}	0.171	15.08
Dielectric constant (ε^0)	ε	ε_{33}^η	4.40	663.2
3rd Order elastic (GPa)	C	C_{333}^E	-300.0×10^9	
3nd Order piezoelectric (ε^0 m/V)	g	$\frac{1}{2}e_{ijklm}$	-1.31	
Electrostrictive (ε^0 F/m)	Q	f_{333}	-4.40	
3rd Order dielectric (F/V)	η	ε_{333}^η	$O(-3.5 \times 10^{-17})$	
Density (kg/m^3)	ρ		2651	7500

Restricting the deformation to 1D, neglecting the Maxwell stress and body force terms, the final weak form expression is given by

$$\int_0^t (-T(1+u_{,Z})v_{,Z} - \Pi\Phi_{,Z} + \rho_0 \ddot{u}v)dX \quad + Tx_{,Z} v \mid_0^t + \Pi\Phi \mid_0^t \ = 0 \quad . \tag{20}$$

Equation (20) was solved for the 1D case using the COMSOL Multiphysics analysis software[25]. The generalized α method was used for the time dependent solver. As this solver is only A-stable, it exhibits spurious high frequency ringing when subjected to a step loading[26]. To address this issue, artificial Rayleigh damping was incorporated by adding an additional weak term in the form[27]

$$d_R = - \int v_R a \frac{\partial v}{\partial x} [\frac{\partial}{\partial t} \frac{\partial u}{\partial x}] dX \quad , \tag{21}$$

to the LHS of (20) where v_R is an adjustable parameter used to minimize the oscillations in the solution.

MATERIALS PROPERTIES

From (16) and (17) it is seen that eight material properties are required. Three of these are the typical elastic moduli, a, piezoelectric coupling term, e, and the dielectric constant, ε. However, the remaining five coefficients are not always readily available. Davison and Graham[28] present 3rd order elastic coefficients for several materials. However, they only have 3rd order piezoelectric constants for lithium niobate and quartz. These constants can also be estimated, with appropriate assumptions, using pressure derivatives as discussed by Clayton[22] if such data is available. While this approach could be used, for the purposes of this paper, it is sufficient to recognize that when the response is restricted to 1D that the distinction between different crystal classes is not relevant. Accordingly, the coefficients of X-cut quartz for which these properties have been experimentally determined have been used. The properties are shown in Table I. The linear values for PZT are also shown. Both quartz and PZT were used in verifying the numerical implementation by comparing the predicted response with exact solutions to the linear equations.

RESULTS

The solution of equation (20) was verified by comparing with exact solutions obtained by using a modified Dubner-Abate-Crump (DAC) algorithm for numerically inverting the Laplace transform[29] of the linear piezoelectric equations for a simple disk of unit cross-sectional area and unit thickness. The poling direction was aligned with the axial direction of the disk. In the first verification, a short circuit solution was obtained for a step voltage with the boundary conditions listed in Table II. The voltage

Table II: Boundary Conditions

Location	Step Voltage	Resonance	Step Pressure
$x = 0$	$\phi = H(t), T = 0$	$\phi = 0, T = sin(\omega t)$	$\phi = 0, T = T_0 H(t)$
$x = t$	$\phi = 0, T = 0$	$\phi = 0, T = 0$	$\phi = 0, u = 0$

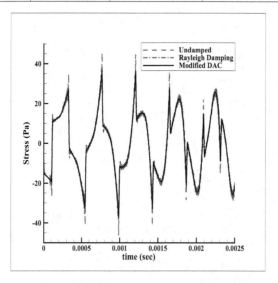

Figure 1: Transient stress history (at x=2.15 mm) in a 4.3 mm thick PZT-4 disk subjected to a Heaviside step voltage.

drop was one volt across the thickness, t. In the second verification, an oscillatory pressure boundary condition was applied. The frequency, ω, was chosen to be close to the axial resonance frequency of the disk.

Figures [1] and [2] show COMSOL numerical results to a step voltage for PZT-4 and quartz, respectively. Figure [1] shows the stress response at the location $x = L/2$ for PZT-4. It is seen that the analytic solution obtained via the modified DAC techniques and the weak form solution compare very favorably. Also, the response shown agrees qualitatively with experimental results obtained by Stuetzer[30] for PZT-4. The effect of adding artificial damping is clearly evident as the Rayleigh damping completely eliminates the spurious oscillations. Similarly, the modified DAC algorithm and COMSOL numerical results for the response of a quartz disk subjected to a unit step voltage are in excellent agreement. In the case of quartz, however, the numerical oscillations due to the solver response to step loadings is more pronounced. While, the inclusion of the numerical damping completely eliminates these oscillations there is a loss of accuracy at the step transitions as a larger value of the damping parameter v_R was required to fully damp the solver oscillations.

Finally, the implementation was verified by comparing the resonance response, Figure [3]. As sinusoidal loading is reasonably smooth, solver oscillations are not induced. It is seen that the numerical and analytical solutions are again in excellent agreement. The effect of the numerical damping is also clearly

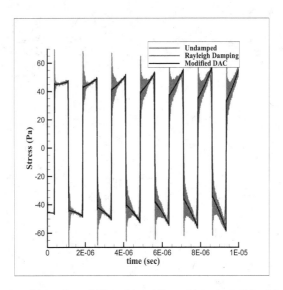

Figure 2: Transient stress history (at x=2.15 mm) in a 4.3 mm thick quartz disk subjected to a Heaviside step voltage

seen in the reduction in the peak stresses of each loading cycle.

With the numerical solution technique verified by comparison with solutions obtained with the modified DAC algorithm solutions, several cases were studied corresponding to a step pressure load applied at the $x = 0$ endpoint location. The boundary conditions are given the third column of Table II. The stress response is normalized in order to highlight the effect of pressure on the response. The 1 GPa-linear curve represents the exact solution of the linear piezoelectricity equations to a 1 GPa step pressure. The numerical solution of the nonlinear equations is indistinguishable for this load and is therefore not shown in Figure [4]. It is seen that the 5 GPa-compression result is also almost the same as the 1 GPa-linear result. The principal differences are that the normalized peak stress is approximately -2.25 compared with the peak value of -2.0 for the linear result. The wave speed is nearly the same for these loads as indicated by the coincidence of the step changes in the response as the stress wave reflects from the boundaries. Similarly, the 5 GPa-tension and 10 GPa-compression curves also retain the same basic form as the linear response. However, the wave speed for these two cases, while different from the linear response, are the same. This indicates that the wave speed is not symmetric with respect to the loading level. Furthermore, the wave speed is slower for the 10 GPa compressive load than the linear response. This is in contrast with the linear theory which predicts that the wave speed should increase with increasing compression load level.

An expression for the effective wave speed, c, can be obtained by substituting (16), (17) and (19) into the 1D form of equation of motion, (7). This leads to

$$c^2 = \frac{(a + C\Gamma - gE)(1 + u_{,z})^2 + (a\Gamma - eE + \frac{1}{2}C\Gamma^2 - \frac{1}{2}QE^2 - g\Gamma E)}{\rho_0} \quad . \tag{22}$$

If terms arising in (22) related to nonlinear affects are dropped, then the expression for c^2 agrees with

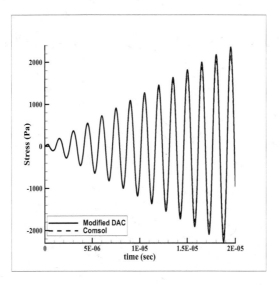

Figure 3: Resonance response (at x=2.15 mm) in a 4.3 mm thick quartz disk subjected to an harmonic voltage. $\omega = 666$ kHz

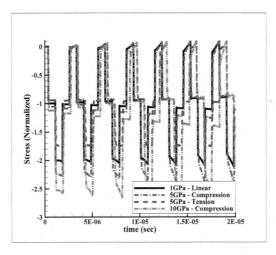

Figure 4: Transient stress histories (at x=2.15 mm) in a 4.3 mm thick quartz disk subjected to various Heaviside step stress loadings

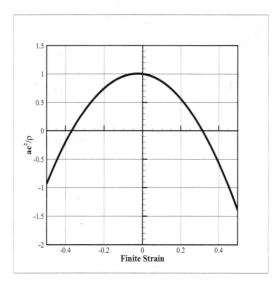

Figure 5: Normalized wave speed squared ac^2/ρ vs. finite strain

classical results[31]. Figure [5] plots c^2 normalized by the linear elastic bar velocity versus the finite strain measure Γ. At $\Gamma = 0$, the wavespeed is the same as classical elasticity. For slightly compressive strains, $\Gamma < -2.67\%$, the wave speed increases. However, for higher strain levels, the wave speed decreases. Also, for large finite strains, in either compression or tension, the wave speed is imaginary which could represent the transition to standing waves. Whether this is correct or simply represents a constraint on the range of validity of the theory is currently being investigated.

CONCLUSIONS

A second order theory for the behavior of piezoelectric materials has been presented. It has been shown that the presented theory reproduces to the same results as linear piezoelectricity for small strains. The effects of the nonlinear terms have been shown by computing the response of quartz to various levels of pressure which are realistic for shock loading. In these cases, the theory predicts that for large enough compressive or tensile pressures that the wave speed will decrease. This is in contrast to linear piezoelectricity where the wave speed always increases with increasing compressive stresses. Finally, the second order theory predicts that governing equations will change behavior at large absolute values of the finite strain. Whether this represents a real phenomena or is simply a restriction on the range of applicability of the theory is currently being investigated.

REFERENCES

[1]W. Cady, Piezoelectricity: An Introduction to the Theory and Applications of Electromechanical Phenomena in Crystals, Dover 1964.

[2]T. Chen, Further correspondences between plane piezoelectricity and generalized plane strain in elasticity, Proc. R. Soc. Lond., A454, 873-884 (1971).

[3]S. Chizhikov, N. Sorokin and V. Petrakov, The elastoelectric effect in the non-centrosymmetric crystals, Ferroelectric, 41, 9-25 (1982).

[4]G. Maugin, Continuum Mechanics of Electromagnetic Solids, North-Holland, (1988).

[5]F. Ashida, and T. Tauchert, An inverse problem for determination of transient surface temperature from piezoelectric sensor measurement, J. Appl. Mech., 65, 367-373 (1998).

[6]D. Chandrasekharaiah, A generalized linear thermo-elasticity theory for piezoelectric media, Acta Mech., 71, 39-49 (1998).

[7]D. Nelson, Theory of nonlinear electroacoustics of dielectric piezoelectric crystals, J. Acoust. Soc. Am., 63, 1738-1748 (1978).

[8]R. Toupin, A dynamical theory of elastic dielectrics, Int. J. Eng. Sci. 21 101-126 (1983).

[9]H. Tiersten, Electroelastic interactions and the piezoelectric equations, J. Acoust. Soc. Am., 70, 1567-15768 (1981).

[10]D. Norwood, M. Shuart, and C. Herakovich, Geometrically nonlinear analysis of interlaminar stresses in unsymmetrically laminate plate subject to inplane mechanical loading, Proceeding of the AIAA/ASME /ASCE/AHS/ACS 32nd Structures, Structural Dynamics and Materials Conference, (AIAA, Washington DC), 938-955 (1991).

[11]G. Kulkami and S. Hanagud, Modeling issues in the vibration control with piezoceramic actuators, Smart Structures and Materials, ed GK, (1991).

[12]P. Pai, A. Nayfeh, and K. Oh, A nonlinear theory of laminated piezoelectric plates, Proceeding of the 33rd AIAA/ASME/ASCE/AHS/ASC Structures, Structural Dynamics and Materials Conference, (AIAA, Washington, DC) AIAA-92-2407-CP, 577-585 (1992).

[13]S. Joshi, Nonlinear constitutive relations for piezoceramic materials, Smart. Mater. Struct, 1, 80-83 (1992).

[14]H. Tiersten, Electroelastic equations for electroded thin plates subjected to large driving voltages J. Appl. Phys. 74 3389-3393 (1993).

[15]J. Yang, and R. Batra, A second-order theory for piezoelectric materials, J. Acoust. Soc. Am., 97 280-288 (1995).

[16]W. Feng, G. Gazonas, D. Hopkins, E. Pan, A second-order theory for piezoelectricity with 6mm and m3 crystal classes, Smart. Mater. Struct, (to appear).

[17]E. Kiral, and A. Eringen, Constitutive equations of nonlinear electromagnetic-elastic crystals, Springer-Verlag (1990).

[18]L. Malvern, Introduction to the Mechanics of Continuous Media. Prentice-Hall, Inc (1969).

[19]Y. Pao, Electromagnetic forces in deformable continua, Mechanics Today vol IV, Pergamon Press, 209-306 (1978).

[20]M. Lax, and D. Nelson, Maxwell equations in material form, Physical Review B, 13, 4, 1777-1784 (1976).

[21]A. R. Dorfmann, R. W. Odgen, Nonlinear electroelasticity, Acta Mechanica, 174, 3-4, 167-183 (2000).

[22]J. Clayton, A non-linear model for elastic dielectric crystals with mobile vacancies, Int. Jour. of Non-Lin Mech., 44, 6, 675-688 (2009).

[23]T. Zielinski, Fundamentals of multiphysics modelling of piezo-poro-elastic structures, Arch. Mech., 62, 5, 343-378 (2010).

[24]W. Feng, E. Pan, X Wang, G. Gazonas, A second-order theory for magnetoelectrostatic materials with transverse isotropy, Smart Mater Struct, 18 (2009).

[25]COMSOL Multiphysics User's Guide, COMSOL AB (2008).

[26]R. Piche, Numerical Experiments with the Modified Rosenbrock Algorithm for Structural Dynamics, Report 17, Tampere (1995).

[27]T. Hughes, The Finite Element Method: Linear Static and Dynamic Finite Element Analysis, Prentice-Hall (1987).

[28] L. Davison and R. Graham, Shock Compression of Solids, Physics Reports (Review Section of Physics Letters), 55, No. 4, 255-379 (1979).

[29] R. Laverty, and G. Gazonas, An improvement to the Fourier series method for inversion of Laplace transforms applied to elastic and viscoelastic waves, Int. J. Comp. Meth., Vol. 3, (1), 57-69 2006.

[30] O. Stuetzer, Multiple Reflections in a Free Piezoelectric Plate, J. Acoust. Soc. Am., 42, 502-508 (1964).

[31] P. Lynse, One-Dimensional Theory of Polarization by Shock Waves: Application to Quartz Gauges, J Appl. Phys, 43, 2, 425-431 (1972).

IMPACT OF MATERIAL AND ARCHITECTURE MODEL PARAMETERS ON THE FAILURE OF WOVEN CMCS VIA THE MULTISCALE GENERALIZED METHOD OF CELLS

Kuang Liu and Aditi Chattopadhyay
Arizona State University
Tempe, AZ 85287

Steven M. Arnold
NASA Glenn Research Center
Cleveland, OH 44135

ABSTRACT

It is well known that failure of a material is a locally driven event. In the case of ceramic matrix composites (CMCs), significant variations in the microstructure of the composite exist and their significance on both deformation and life response need to be assessed. Examples of these variations include changes in the fiber tow shape, void content within tows, tow shifting/nesting and voids within and between tows. In the present work, the effects of many of these architectural parameters and material scatter of woven ceramic composite properties at the macroscale (woven RUC) will be studied to assess their sensitivity. The recently developed Multiscale Generalized Method of Cells methodology is used to determine the overall deformation response, proportional elastic limit (first matrix cracking), and failure under tensile loading conditions. The macroscale responses investigated illustrate the effect of architectural and material parameters on a single RUC representing a five harness satin weave fabric. Results shows that the most critical architectural parameter is the weave void content with other parameters being less in severity. Variation of the matrix material properties was also studied to illustrate the influence of the material variability on the overall features of the composite stress-strain response.

INTRODUCTION

Multiscale modeling has been applied to both laminated and woven composites in the past. Although nomenclature in the literature varies, typically a multiscale modeling analysis will follow length scales shown in Figure 1 for continuum modeling. These scales, progressing from left to right in Fig. 1, are the microscale (constituent level; fiber, matrix, interface), the mesoscale (tow), the macroscale (repeating woven unit cell), and the global/structural scale. Traditionally, one traverses (transcends (moves right) or descends (moves left)) these scales via homogenization and localization techniques, respectively (Fig. 1 and 2a); where a homogenization technique provides the properties or response of a "**structure**" (higher level) given the properties or response of the structure's "**constituents**" (lower scale). Conversely, localization techniques provide the response of the constituents given the response of the structure. Figure 2b illustrates the interaction of homogenization and localization techniques, in that during a multiscale analysis, a particular stage in the analysis procedure can function on both levels simultaneously. For example, during the process of homogenizing the stages represented by X and Y to obtain properties for the stage represented by V, X and Y should be viewed as the constituent level while V is on the structure level. However, during the process of homogenizing V and W to obtain properties for U, V is now on the constituent level (as is W). Obviously, the ability to homogenize and localize accurately requires a sophisticated theory that relates the geometric and material characteristics of structure and constituents.

With the recent development of multiscale generalize method of cells (MSGMC), one can now ascertain the influence of architectural parameters, such as volume fraction, weave geometry, tow geometry, etc., at each associated length scale, for composites, particularly woven and braided composites. This enables the determination of which effect/parameter, at a given length scale, is impactful/relevant at higher length scales. For example, matrix elastic modulus is a microscale effect, changing this value will have a direct effect at the next largest length scale (e.g., mesoscale), but its effect at the macro or structural scale cannot necessarily be assumed. Similarly the tow fiber volume fraction, which is a mesoscale effect, should have a direct impact on the response at the macroscale, yet its effect at the global scale is difficult to deduce a priori. Furthermore, experimental investigations have shown that in typical composite (particularly woven) materials there exist significant variations in the meso and macroscale architectural features. Yet most analyses performed (until now) assume an idealized or pristine material and architecture at every length scale. Such an assumption was required, up until now, to avoid the computationally exhaustive multiscale modeling of every minute variation in architecture at every length scale, via the finite element method. The objective of this paper is to perform a preliminarily yet truly multiscale[1] investigation to study lower length scale effects and determine their significance at the macro and structural scale.

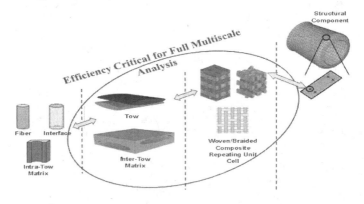

Figure 1 Illustration of associated levels scales for woven/braided composite analysis.

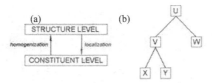

Figure 2 (a) Homogenization provides the ability to determine structure level properties from constituent level properties while localization provides the ability to determine constituent level responses from structure level results. (b) Example tree diagram.

---

[1] Here the term multiscale refers to an analysis in which at least 3 levels of scales are accounted for, wherein at least two homogenizations/localizations are required.

MULTISCALE GENERALIZED METHOD OF CELLS

Overview

Analysis of woven fabric composites can be generalized into several relevant length scales (from largest to smallest): macro, meso, and micro. The macroscale weave refers to the RUC of the weave, for a 5-harness satin fabric, see Figure 3. The mesoscale represents the RUC of the fiber tow, this represents a bundle of fibers (typically 700-1000 for ceramic matrix composites). The smallest length scale is the microscale, which represents the fundamental constituent materials, such as the monofilament fiber and matrix itself.

This multiscale analysis uses the recently developed Multiscale Generalized Method of Cells (MSGMC) methodology[1]. At the macroscale, each fabric composite (e.g., plain or 5HS, see Figure 3) was discretized into N_A, N_B, N_Γ subcells, wherein for example a subcell used to represent a fiber tow, is further represented at the mesoscale by N_β, N_γ subcells, wherein each of these, are represented by the constitutive properties of the fiber and matrix at the microscale. This recursive methodology (wherein the generalized method of cells (GMC), see Paley and Aboudi[2] and Aboudi[3], is called within GMC) is shown in Figure 4, and can be accounted for by attaching the superscript $bg\text{ABT}\beta\gamma$ to each. There are several architectural parameters at the meso, macro, and structural level required to fully define the discretized subcell geometries. At the mesoscale, both tow volume fraction and tow packing are required, while at the macroscale, weave architecture, weave volume fraction, tow aspect ratio and ply nesting are required. At the structural level, the spatial distribution of the macroscale RUCs are required, i.e., uniform – each subcell is associated with the same macroscale RUC or random – subcells are associated with a uniform distribution of macroscale RUCs. It has been of recent interest to study the effects of these parameters and understand what the driving factors for both elastic and inelastic response are.

Figure 3 Five Harness Sating Macroscale RUC.

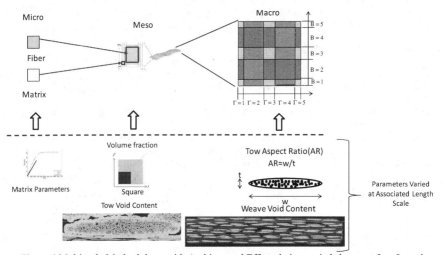

Figure 4 Multiscale Methodology with Architectural Effects being varied shown at four Length Scales considered.

Microscale (Constitutive Modeling)

The Multiscale Generalized Method of Cells (MSGMC) is used to represents the woven fabric composite starting with its constituent materials, i.e. the fiber (monofilament) and matrix and progress up the various length scales. The microscale is the only length scale where explicit constitutive models are applied to the various phases (e.g. fiber and matrix). Stress states and tangent moduli for larger length scales are determined through the Generalized Method of Cells (GMC) triply-periodic homogenization procedure developed by Aboudi[3]. The monofilament fibers are modeled using a linear elastic relationship, i.e. Hooke's Law, and the matrix material is represented by a damage mechanics type relationship based on a tangent modulus relationship. Details for the damage model can be found in the following sections. All microscale constituent parameters (i.e., modulus, failure strength, etc.) were assumed to be *deterministic* in this analysis. The stresses in any subcell in the microscale can be determined from the following equation. The stiffness $C^{bg\mathrm{AB}\Gamma\beta\gamma}$ is determined from the given material parameters and the strains $\varepsilon^{bg\mathrm{AB}\Gamma\beta\gamma}$ are determined from localization from the mesoscale. This is possible through a concentration matrix, A, determined by GMC, which is a function of the subcell geometry and stiffness matrix.

$$\sigma^{bg\mathrm{AB}\Gamma\beta\gamma} = D^{bg\mathrm{AB}\Gamma\beta\gamma} C^{bg\mathrm{AB}\Gamma\beta\gamma} \varepsilon^{bg\mathrm{AB}\Gamma\beta\gamma} \tag{1}$$

$$\varepsilon^{bg\mathrm{AB}\Gamma\beta\gamma} = A^{bg\mathrm{AB}\Gamma\beta\gamma} \varepsilon^{bg\mathrm{AB}\Gamma} \tag{2}$$

Mesoscale (Tow)

The mesoscale is used to represent the periodic structure of a fiber tow. At the mesoscale, there are two significant architectural parameters: fiber packing and tow volume fraction. Both of these parameters govern the mesoscale subcell geometries. The response of the mesoscale is subject to these parameters as well as the material variation at the microscale. The continuous fiber tows are assumed to be represented by a doubly-periodic RUC of dimensions h by l consisting of constituents from the microscale. An example of such an RUC discretized for GMC is shown in Figure 4, where the inner region (shown in grey) denotes the fiber tow and the outer region (shown in white) is the matrix. The RUC is discretized in such a manner that it is composed of $N_\beta \times N_\gamma$ rectangular subcells, with each subcell having dimensions $h_\beta \times l_\gamma$. From this point forward, superscripts with lowercase Greek letters denote a specific subcell at the microscale, superscripts of uppercase Greek letters denote a specific subcell at the macroscale and superscripts with lowercase Roman letters denotes macroscale variables. Fiber tow packing and volume fraction typically govern the architecture of the mesoscale RUC but must be in accordance with the previously described RUC microstructural parameters. The resulting stress in the fiber tow can be determined from the GMC homogenization process, where in GMC, the current stress and current tangent moduli of a particular fiber tow at a point are determined through a volume averaging integral over the repeating unit cell. This process is represented by the summation in the following equations, producing the first homogenization in the multiscale modeling framework. In these equations, σ denotes the Cauchy true stress, A denotes the strain concentration matrix, and C denotes the stiffness matrix[2,3]. The microscale subcell stresses and tangent moduli needed to complete the summation are determined through the applied constitutive models for each constituent based on their current strain state. The mesoscale strains, which are used as the boundary conditions for the GMC analysis, are determined from the through thickness (tt) homogenization at the macroscale. The subscripts tt in the concentration matrix in Eq. 5 denote the 2nd portion of the two step homogenization process discussed later.

$$\sigma^{bg\,AB\Gamma} = \left(\frac{1}{hl} \sum_{\beta=1}^{N_\beta} \sum_{\gamma=1}^{N_\gamma} \sigma^{\beta\gamma} h_\beta l_\gamma \right)^{bg\,AB\Gamma} \tag{3}$$

$$C^{bg\,AB\Gamma} = \left(\frac{1}{hl} \sum_{\beta=1}^{n_\beta} \sum_{\gamma=1}^{n_\gamma} D^{\beta\gamma} C^{\beta\gamma} A^{\beta\gamma} h^\beta l^\gamma \right)^{bg\,AB\Gamma} \tag{4}$$

$$\varepsilon^{bg\,AB\Gamma} = A_{tt}^{bg\,AB\Gamma} \varepsilon^{bg\,B\Gamma} \tag{5}$$

$$\sigma^{bg\,AB\Gamma} = C^{bg\,ABT} \varepsilon^{bg\,ABT} \tag{6}$$

Macroscale (Weave)

At the macroscale the RUC for the weave fabric is modeled. At this scale, the architecture is governed by the overall volume fraction, tow geometry (aspect ratio, width and thickness), and overall fabric thickness, wherein the subcell "constituent" response is obviously dependent on the mesoscale and microscale responses. The weave requires a triply-periodic RUC representation, of size $D \times H \times L$ and discretized into $N_A \times N_B \times N_\Gamma$ parallelepiped subcells, with each subcell having dimensions $D_A \times H_B \times L_\Gamma$. At this length scale, a two step homogenization procedure was employed to determine the stiffness and macroscale stresses. This is to overcome the lack of shear coupling inherent to the GMC formulation[5]. The first step involves a through thickness homogenization and the second step is an in-plane homogenization, where subscripts tt and ip denote through thickness and in-plane respectively. Details for the subcell geometry and RUC information can be found in Refs. 5 and 6.

<u>Through thickness homogenization</u>

$$\sigma^{bg\mathrm{B}\Gamma} = \left(\frac{1}{D}\sum_{\mathrm{A}=1}^{N_\mathrm{A}}\sigma^\mathrm{A}D_\mathrm{A}\right)^{bg\mathrm{B}\Gamma} \tag{7}$$

$$C^{bg\mathrm{B}\Gamma} = \left(\frac{1}{D}\sum_{\mathrm{A}=1}^{N_\mathrm{A}}A_{tt}^\mathrm{A}C^\mathrm{A}D_\mathrm{A}\right)^{bg\mathrm{B}\Gamma} \tag{8}$$

$$\varepsilon^{bg\mathrm{B}\Gamma} = A_{ip}^{bg\mathrm{B}\Gamma}\varepsilon^{bg} \tag{9}$$

$$\sigma^{bg\mathrm{B}\Gamma} = C^{bg\mathrm{B}\mathrm{T}}\varepsilon^{bg\mathrm{B}\mathrm{T}} \tag{10}$$

<u>In-plane homogenization</u>

$$\sigma^{bg} = \left(\frac{1}{HL}\sum_{\mathrm{B}=1}^{N_\beta}\sum_{\Gamma=1}^{N_\Gamma}\sigma^{\mathrm{B}\Gamma}H_\mathrm{B}L_\mathrm{T}\right)^{bg} \tag{11}$$

$$C^{bg} = \left(\frac{1}{HL}\sum_{\mathrm{B}=1}^{N_\beta}\sum_{\Gamma=1}^{N_\Gamma}A_{ip}^{\mathrm{B}\Gamma}C^{\mathrm{B}\Gamma}H_\mathrm{B}L_\mathrm{T}\right)^{bg} \tag{12}$$

$$\varepsilon^{bg} = A^{bg}\varepsilon \tag{13}$$

$$\sigma^{bg} = C^{bg}\varepsilon^{bg} \tag{14}$$

MODELING CERAMIC MATRIX COMPOSITES WITH MSGMC

Weave Repeating Unit Cell

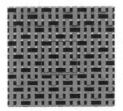

Figure 5 Five Harness Satin Repeating Unit Cell

For this particular study, a five harness satin weave is considered. In this idealization of the architecture, the repeating unit cell is assumed to be representative of the entire structure. A picture of the fabric with the repeating unit cell outlined is shown in Figure 6. To create a RUC suitable for analysis, the weave is discretized into several subvolume cells. There are two types of materials comprising all the subcells: fiber tows and interweave matrix. This final 3D discretization is shown in

Figure 6, in addition with the multiscale analyses of the voids, tows and tow voids. In the figure, fiber tows are indicated through the lined subcells. The lines indicate the direction of orientation. The blank subcells represent the interweave matrix. This results in a 10x10x4 sized RUC of dimensions shown in Eq. 15. In this equation w is the tow width and delta can be determined from $V_f = \dfrac{w}{(w+\delta)}V_{f_{tow}}$. The proper overall fiber volume fraction and tow width is enforced by back calculating the tow spacing. Due to the chemical vapor infusion process used to manufacture the woven fabric composites, there exists high levels of porosity, as shown in Figure 4, that cannot be neglected.

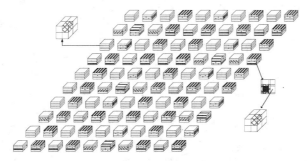

Figure 6 Discretized 5-Harness Satin Subcell Configuration

Voids are accounted for in the RUC in one of three ways: 1) void content is neglected; 2) voids are assumed to be evenly distributed through the weave; or 3) voids are localized to critical areas determined from optical inspection (microscopy). Figure 7 illustrates the three types of void modeling at the macroscale. The first figure shows no voids accounted for anywhere, the second figure depicts voids evenly distributed in the weave matrix (yellow), and the third figure shows high density void regions in red and low density void regions in blue. The voids are accounted for at a smaller length scale by analyzing a separate RUC homogenizing those properties. This is done for two primary reasons. First, modeling of voids in GMC will tend "eliminate" an entire row and column due to the constant strain field assumptions within a subcell. By performing a separate analysis, this effect is dampened. Secondly, this allows for a faster, more accurate representation of void shape and distribution then explicitly modeling voids at this length scale.

$$D = \{t/4, t/4, t/4, t/4\}$$

$$H = \{\delta, w, \delta, w, \delta, w, \delta, w, \delta, w\}$$

$$L = \{\delta, w, \delta, w, \delta, w, \delta, w, \delta, w\} \tag{15}$$

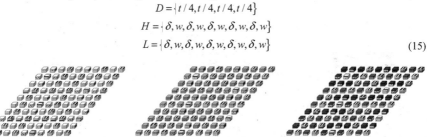

Figure 7 Three Types of Void Distributions

Tow Repeating Unit Cell

The fiber tow bundles are modeled using a doubly periodic (continuously reinforced) 4x4 repeating unit cell consisting of three materials: fiber, fiber coating/interface, and matrix. In Figure 8, the black denotes the fiber, the hatched area represents the interface, and white represents the matrix.

At this level there are also voids due to the CVI process. The voids at this level appear to be more evenly distributed than at the weave level and thus are represented by evenly distributing the void content in the tow areas (see tow insert in Figure 4). Each matrix subcell in the RUC will call a separate void analysis, just as described in the weave RUC section. For each fiber tow bundle, the orientation is carefully computed such that the undulation is properly accounted for and the failure criteria can be applied in the local coordinate system.

Figure 8 Fiber Tow Bundle RUC

Void Modeling

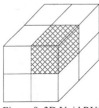

Figure 9 3D Void RUC

Voids are modeled through computation of a triply periodic (discontinuously reinforced) 2x2x2 RUC as shown in Figure 9. The hatched subcell represents the void portion while the white represents the matrix. The relative size of the void cell is what determines the overall void content in both the fiber tow bundles and the weave. As mentioned previously, modeling of voids as a separate GMC analysis has many advantages. The overall RUC of the weave will remain constant regardless of the shape and distribution of the voids, i.e. no rediscretization is required. Consequently, the void location, quantity, and geometry can be quickly changed. Lastly, the strength and stiffness degradations and stress concentrations can be captured through GMC without reducing the accuracy of the analysis at the macroscale.

CONSTITUTIVE AND FAILURE MODELING

Matrix Damage Modeling

The matrix material, in both the weave and tow, is modeled through a damage mechanics constitutive model based on the first invariant of the stress tensor. This constitutive model represents the cracks and brittle failure often seen in these CMCs. A damage scalar, ϕ, which varies between zero (no damage) and one (complete failure/damage), scales the elastic portion of the stiffness tensor and is employed directly in the stress strain relationship, shown in Eq. 16.

$$\sigma = (1-\phi)C\varepsilon \tag{16}$$

To determine the damage scalar, first a damage rule is defined as:

$$f = 3\varepsilon_{eq}nK - \sigma_{eq} = 0 \tag{17}$$

In this potential, n represents the damaged normalized secant modulus and K represents the instantaneous tangent bulk modulus. This potential uses an equivalent stress and strain as defined by the invariants of the respective tensors. This is shown in the following equations.

$$\sigma_{eq} = (\sigma_{11} + \sigma_{22} + \sigma_{33})/3 = I_1(\sigma)$$

$$\varepsilon_{eq} = (\varepsilon_{11} + \varepsilon_{22} + \varepsilon_{33})/3 = I_1(\varepsilon) \tag{18}$$

The damage rule in (17) is only applicable after a critical stress criteria has been reach, i.e. it is only valid when $\sigma_{eq} > \sigma_{dam}$. Equation 16 can be rewritten in incremental form with $i+1$ denoting the next increment $(\varepsilon^{i+1} = \varepsilon^i + \delta\varepsilon^{i+1})$.

$$f = n3K^i\delta\varepsilon_{eq}^{i+1} - \delta\sigma_{eq}^{i+1} = 0 \tag{19}$$

This can be converted to a strain based function by substituting the following relationship in for the stress increment:

$$\delta\sigma_{eq}^{i+1} = 3\left(K^{i+1}\delta\varepsilon_{eq}^{i+1} + \left(K^{i+1} - K^i\right)\varepsilon_{eq}^{i+1}\right) \tag{20}$$

Resulting in:

$$nK^0\delta\varepsilon_{eq}^{i+1} - \left(K^{i+1}\delta\varepsilon_{eq}^{i+1} + \left(K^{i+1} - K^i\right)\varepsilon_{eq}^{i+1}\right) = 0 \tag{21}$$

where K^0 represents the initial bulk modulus. The instantaneous tangent bulk modulus can be related back to the damage scalar through:

$$K^{i+1} = \left(1 - \phi^{i+1}\right)K^0 \tag{22}$$

Substitution of (22) into (21) and simplification yields a formulation for the damage scalar.

$$1 - \phi^{i+1} = \lambda^{i+1} = K^0 \frac{n\delta\varepsilon_{eq}^{i+1} + \lambda^i\varepsilon_{eq}^{i+1}}{\left(\delta\varepsilon_{eq}^{i+1} + \varepsilon_{eq}^{i+1}\right)} \tag{23}$$

Where the initial value, ϕ^0, is zero.

Fiber Failure Modeling

The fiber is model through a simple linear elastic constitutive model, but employs a Hashin type failure criterion. This criterion determines the catastrophic failure of the fiber based on the axial and shear strengths. When the failure criterion exceeds 1, the stiffness matrix is degraded to a minimal value. A key assumption made in this analysis is that the fiber interface will fail simultaneously with the fiber and does not present its own failure modes. The failure stress levels presented later are an in-situ failure stress considering the interface.

$$f = \frac{\sigma_{11}^2}{\sigma_{axial}^2} + \frac{1}{\tau_{axial}^2}\left(\sigma_{13}^2 + \sigma_{12}^2\right) \tag{24}$$

RESULTS

Problem Description

For this study, a five harness satin weave with a CVI-SiC matrix and iBN-Sylramic fiber (silicon carbide fiber coated with boron nitride) were chosen, due to the availability of experimental data for correlation. An approximate overall fiber volume fraction of 36% was determined along with a tow width of 10mm and thickness of 2.5mm, see actual micrograph inserts in Fig. 4. The properties and necessary material parameters are displayed in the table below, most properties were determined between either published values or discussions with colleagues.

Table 1. Fiber Properties

Name	iBN-Sylramic
Modulus	400 GPa
Poisson's Ratio	0.2
Axial Strength	2.2 GPa
Shear Strength	900 MPa

Table 2. Matrix Properties

Name	CVI-SiC
Modulus	420 GPa
Poisson's Ratio	0.2
σ_{dam}	180 MPa
n	0.04

Table 3. Interface Properties

Name	Boron Nitride
Modulus	22 GPa
Poisson's Ratio	0.22

Table 4. Weave Properties

Type	5HS
Fiber Volume Fraction	36%
Tow Volume Fraction	78%
Tow Width	10mm
Tow Spacing	2.78mm
Thickness	2.5mm
Matrix	CVI-SiC

Table 5. Tow Properties

Tow Fiber Volume Fraction	46%
Tow Packing Structure	Square
Fiber	IBN-Sylramic
Matrix	CVI-SiC
Interface	BN

Typical Results

A typical response curve of an experimental, on-axis, tensile test is shown in Figure 10, taken from Morscher[7], and is overlaid with a baseline prediction using the localized void model (see Fig. 7c). The predicted response shows good correlation with the experimental curve, approximately capturing the first matrix cracking and failure stress. In Figure 11, the underlying mechanisms causing nonlinearity (which is subtle in some places), are denoted; the four primary events being: inter-tow matrix damage, inter-weave matrix damage (in the low stress and also in the high stress region) and fiber failure. The multiple damage initiation points are due to two reasons. First, different regions of the weave RUC will initiate damage at different times. Secondly, different subcells of a tow within a given region initiate damage at different times. It is useful also to look at the instantaneous secant elastic modulus, which degrades due to matrix damage as shown in **Error! Reference source not found.**. It is easier to understand the degradation effects due to the matrix by directly looking at the stiffness effects. In a typical tensile response curve, there are four significant events that are useful for characterizing the material; these are: 1) initial modulus 2) point of nonlinearity 3) damaged modulus and 4) failure point. The subsequent parametric study will focus on the impact that material and architectural parameters have on these four significant measures.

Furthermore, it is critical to understand the underlying mechanisms governing these events. In the case of the initial modulus, it is clear that the individual constituents' stiffness matrices and the weave architecture are primary drivers, along with possible microcracking of the matrix constituent. The fact that some damage occurs before the first major point of nonlinearity, is substantiated by the experimental acoustic emission results in Ref. 7. Similarly, the model attributes this initial cracking to damage in the intertow matrix and to damage in the high void density region of the interweave matrix (known as the high stressed region). The second event (i.e., the first major point of nonlinearity) occurs at approximately 0.075% strain, for the CMC examined, is said to be "first matrix cracking". This point is taken to reflect a significant crack (or coalescence of microcracking) occurring in either the tow or weave matrix. Correlating model results to that of the typical response (see Fig. 7), the model predicts that cracking occurs in both the tow and weave, at first matrix cracking. Thirdly, the slope of the post first matrix cracking curve (damage modulus) is determined by the response of the matrix material (i.e., the behavior after damage initiation) and corresponding constitutive model and weave architecture. Again, the experimental acoustic emission results (of Ref 7) also show some damage gradually occurring after first matrix cracking within this region. This is most likely a combination of all previous damage growing as well as the onset of new damage in the high stressed regions. This damage progression continues with continuous local stress redistribution until the final failure point is determined by the failure strength of the fiber. Note, although not considered here, MSGMC can incorporate statistical fiber breakage by modeling multiple fibers within the Tow RUC.

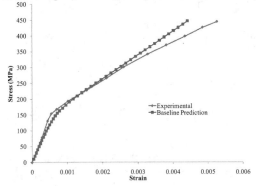

Figure 10 Typical Experimental Response Curve[7]

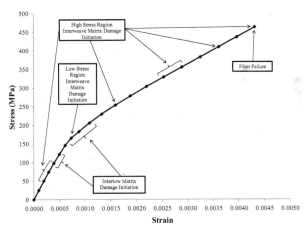

Figure 11 Typical Simulated Response Curve

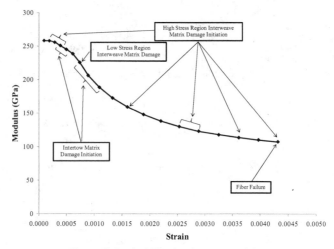

Figure 12 Typical Simulated Secant Modulus

Effects of Material Properties

To understand the influence of constituent material variation; three of the parameters for the constitutive model of the matrix material (i.e., the modulus, n, and σ_{dam}) were varied to understand their effect on the overall macro response. Note, the properties were varied a significant amount from the baseline so that their effect could be clearly seen. For example, the modulus was increased by 50%, n by 200% and σ_{dam} was increased by 100%. Considering the results in Figure 11, one would expect that changing the matrix modulus should correspond to changing the initial weave modulus and post first matrix cracking modulus. This is in agreement with the results shown in Figure 13.. In addition to the stiffness changing, the onset of first matrix cracking is also affected; resulting in a higher stress level (approximately 10%) and lower strain to failure (approximately 10%). Next changing only the parameter n from that of the baseline, one would expect the post first matrix cracking modulus to be primarily impacted, as verified in Figure 13, with a corresponding change in failure stress (e.g., increased 10%) and failure strain (e.g., decreased 12%). Finally, increasing σ_{dam} caused the first matrix cracking onset to be delayed (approximately 110 MPa) resulting in higher overall failure stresses and a lower failure strain level of 0.0031. Note that the modulus and post first matrix cracking modulus are nearly unchanged, in this case. The fiber failure stresses were not varied due to the straightforward effect if linearly increasing the failure point.

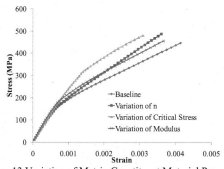

Figure 13 Variation of Matrix Constituent Material Properties

Effects of Architecture

To study the effects of architectural variation on the macroscale response, a full factorial set of numerical simulations were conducted. The parameters varied are shown in Table 6 and depicted in Figure 4. The three tow architectural parameters varied were: a) tow fiber volume fraction, b) tow aspect ratio, and c) tow void volume fraction. In addition, three weave void location cases were examined to illustrate the influence of voids due to manufacturing as well. All other parameters in the analysis were kept constant. There are additional architecture effects that the authors did not investigate that could possibly have an effect. These include interplay nesting, fiber packing structure, coating thickness, fiber tow shifting and others. Future work will determine which parameters are the most significant. The tow fiber volume fraction and void volume fraction are both considered a mesoscale effect because their geometrical properties are involved in the mesoscale concentration matrix (Eq. 2). Whereas, the tow aspect ratio is considered a macroscale property because it is taken into account in the macroscale concentration matrices (Eqs. 5 and 11). The tow volume was varied over a range indicative of the experimental variation: 0.46, 0.48, and 0.50. These three values were chosen based on common experimental values obtained for CMCs. Similarly, realistic tow aspect ratios were chosen, i.e., 8, 10 and 12, where a value of 10 is typical for CMCs and three different fiber void volume fractions were used; 0.01, 0.05, and 0.07.

Table 6 Varied Parameters

Architectural Parameter	Relevant Length Scale	Values
Tow Fiber Volume Fraction $\left(V_{tf}\right)$	Meso	0.46,0.48,0.50
Tow Void Volume Fraction	Meso	0.01,0.05,0.07
Tow Aspect Ratio $\left(AR\right)$	Macro	8,10,12
Weave Void Distribution	Macro	None, Even, Localized

First a plot showing the effects of weave void distribution will be discussed. In Figure 14, there are three line plots each corresponding to a type of analysis for voids discussed earlier. Two cases, no void modeling and evenly distributed voids, both do not capture the correct overall response. The initial modulus is too stiff, and the failure stress levels are too high. This is a result incorrect local failure modes and stress distribution. This demonstrates that an accurate analysis must contain localized void distributions in the correct configuration in order to properly capture the overall deformation and failure response.

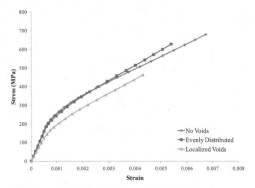

Figure 14 Effects of Weave Void Distribution

The remaining parametric cases were all computed using the localized void model. However, it is important to note that the tow fiber volume fraction and weave void volume fraction are coupled and cannot be decoupled within the analysis, since when the fiber volume fraction within the tow increases, the tow spacing must increase in order to maintain continuity of the overall fiber volume fraction and thickness. This therefore creates a large volume domain for voids to fill, thus increasing the overall void content. Correspondingly, the effect of increasing void content and tow volume fraction are coupled together. The total variation in stress – strain response for all cases are shown in Figure 15. Clearly, the overall response characteristic is very similar, irrespective of the value of the individual parameters, with variations in the initial modulus being relatively minimal and ultimate tensile stress for all practical purposes identical. However, the post first matrix cracking modulus and therefore final failure strain are affected, e.g. the maximum difference being 25%. The configuration providing the stiffest response is composed of a tow volume fraction of 46%, aspect ratio (AR) equal to 12 and tow void fraction of 1%, whereas the most compliant response is generated using a tow volume fraction of 50%, aspect ratio (AR) equal to 8 and tow void fraction of 7%.

In Figures 16 through 18 the various responses are arranged so as to enable identification of parameter sensitivities. Figure 16 shows the effect of tow void content on the overall response; where it can be seen that increasing the void content within the tow causes the response curve to be more compliant with generally an effect of increasing the strain to failure. Figure 17 shows the effect of tow aspect ratio; where increasing the aspect ratio has the effect of stiffening the response curve and lowering the failure stress. Figure 18 displays the influence of tow fiber volume fraction, which appears to be minimal at first glance. Although this trend is possible, as mentioned previously, it is strongly coupled with the overall weave void volume fraction and thus these two effects could be working in opposition to one another. Consequently, it is impossible to deduce from these graphs, the overall effect of tow fiber volume fraction.

Comparing all parameters, the weave void locations, tow void content, tow fiber volume fraction and tow aspect ratio, one can assess the severity of these effects. For example, it is clear from Figure 14, that the location (and shape – not shown here) of voids at the macroscale is a critical driving parameter. This far outweighs all other parameters. Similarly, the effect of interply tow nesting could also be a critical/primary driving factor, yet this effect has been left for future work. Besides the weave void content (i.e., location and shape), the tow void content has the strongest effect on post first matrix cracking stiffness and the tow aspect ratio has the strongest effect on failure strain. The tow fiber volume fraction appears to have a minimal effect.

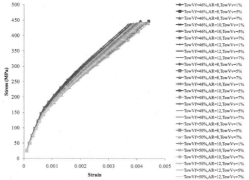

Figure 15 All Simulated Cases

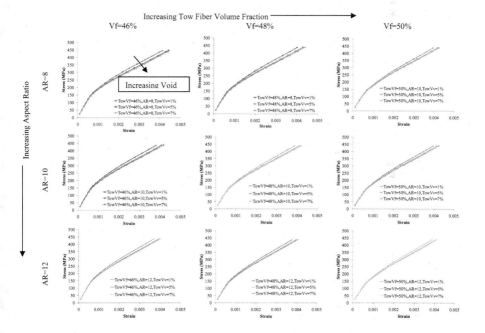

Figure 16 Effects of Tow Void Content

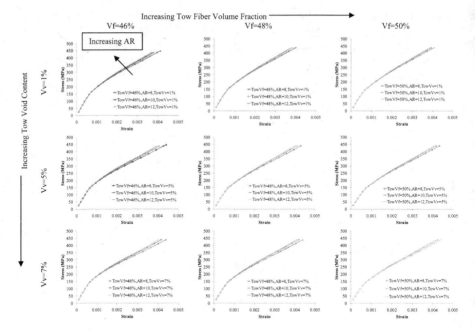

Figure 17 Effects of Aspect Ratio

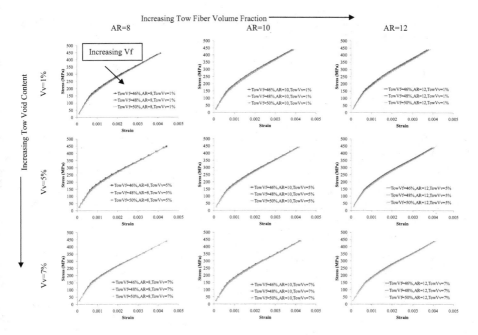

Figure 18 Effects of Tow Volume Fraction

CONCLUSION

This paper presents a detailed investigation into several architectural and material parameter effects on the macroscale deformation response of woven CMCs. The recently implemented Multiscale Generalized Method of Cells methodology was employed to model the nonlinear damage driven response of a five harness satin woven composite fabric, where three separate material scales were considered. At the microscale, the influence of constituent material properties was investigated. At the mesoscale, the tow fiber volume fraction and the void content within a tow were varied; whereas at the macroscale the influence of the tow aspect ratio and weave void content were investigated. For each permutation of these effects, the tensile response was analyzed; wherein the modulus, first matrix cracking, post damage modulus, and ultimate failure strain were examined. Analyzing the macroscale response, it was determined that the weave void content at the macroscale was the most impactful parameter. It is critical that this effect be captured and correctly reflected in the model to ensure accurate deformation and failure response. Second to this, tow void content had the largest effect on the initial and post stiffness and the tow aspect ratio greatly affected the failure strain levels. The general variation was observed to be less than that of a previous polymer matrix composite study and as such one can deduce that the relatively low mismatch in properties within a CMC is the cause of this. Structural scale analysis parameter sensitivity along with tow nesting will be a topic of future work.

REFERENCES

[1] Liu, K. and Chattopahdyay, A. and Bednarcyk, B. and Arnold, S.M., "Efficient Multiscale Modeling Framework For Triaxially Braided Composites using Generalized Method of Cells," Journal of Aerospace Engineering, April 2011.

[2] Paley, M. and Aboudi, J., "Micromechanical Analysis of Composites by the Generalized Method of Cells," Mechanics of Materials 1992;14:127-139.

[3] Aboudi, J., "Micromechanical Analysis of Thermo-Inelastic Multiphase Short-Fiber Composites'", Composite Eng. **5**, 839-850, 1995.

[5] Bednarcyk, B.A. "Modeling Woven Polymer Matrix Composites With MAC/GMC". *NASA/CR.2000-210370.*

[6] Liu, K.C., Hiche, C. and Chattopadhyay, A., "Low Speed Projectile Impact Damage Prediction and Propagation in Woven Composites" 50th AIAA/ASME/ASCE/AHS/ASC Structures, *Proc.* Structural Dynamics and Materials Conference, Palm Springs, CA, May 2009.

[7] Morscher, G. "Stress, matrix cracking, temperature, environment, and life of SiC/SiC woven composites" International Conference on High Temperature Ceramic Matrix Composites, Bayreuth, Germany in September 20-22, 2010.

KINETIC MONTE CARLO SIMULATION OF OXYGEN AND CATION DIFFUSION IN YTTRIA-STABILIZED ZIRCONIA.

Brian Good, Materials and Structures Division, NASA GRC, Cleveland, Ohio.

ABSTRACT

Yttria-stabilized zirconia (YSZ) is of interest to the aerospace community, notably for its application as a thermal barrier coating for turbine engine components. In such an application, diffusion of both oxygen ions and cations is of concern. Oxygen diffusion can lead to deterioration of a coated part, and often necessitates an environmental barrier coating. Cation diffusion in YSZ is much slower than oxygen diffusion. However, such diffusion is a mechanism by which creep takes place, potentially affecting the mechanical integrity and phase stability of the coating. In other applications, the high oxygen diffusivity of YSZ is useful, and makes the material of interest for use as a solid-state electrolyte in fuel cells.

The kinetic Monte Carlo (kMC) method offers a number of advantages compared with the more widely known molecular dynamics simulation method. In particular, kMC is much more efficient for the study of processes, such as diffusion, that involve infrequent events.

We describe the results of kinetic Monte Carlo computer simulations of oxygen and cation diffusion in YSZ. Using diffusive energy barriers from ab initio calculations and from the literature, we present results on the temperature dependence of oxygen and cation diffusivity, and on the dependence of the diffusivities on yttria concentration and oxygen sublattice vacancy concentration. We also present results of the effect on diffusivity of oxygen vacancies in the vicinity of the barrier cations that determine the oxygen diffusion energy barriers.

INTRODUCTION

Zirconia-based materials are of interest for a variety of technological applications. Pure zirconia exists in a monoclinic structure below about 1100C, a tetragonal structure between 1100C and 2300C, and a cubic structure between 2300C and the melting point [1]. Because the high-temperature phases are not stable at room temperature, the utility of pure zirconia for high-temperature applications is limited, especially in applications that involve thermal cycling.

However, the tetragonal and cubic phases can be stabilized via substitutional cation doping, with aliovalent ions such as Y^{3+} or Ca^{2+} replacing Zr^{4+} ions. The cubic phase can be fully stabilized via doping with Y^{3+}, such that the resulting yttria-stabilized zirconia (YSZ) remains in the same phase from room temperature to the melting point. YSZ's thermal stability and low thermal conductivity make it suitable for high-temperature applications such as thermal barrier coatings for turbine engine components.

When zirconia is cation-doped, additional compensating oxygen vacancies are formed so as to maintain overall electrical neutrality. Because oxygen conductivity in these materials occurs via diffusive oxygen ion hopping among vacancy sites, increasing the concentration of such vacancies tends to increase the oxygen diffusivity and ionic conductivity, making such materials of interest for use as oxygen sensors, or as solid electrolytes for fuel cells. However, the oxygen diffusivity does not increase monotonically with Y^{3+} concentration; it increases at low concentrations, but reaches a maximum between 0.08 and 0.15 Y_2O_3, mole fraction, and decreases at higher concentrations [2].

In addition to oxygen diffusion, YSZ also exhibits cation diffusion, which occurs via cation hopping among vacancy sites on the cation sublattice. While such diffusion is orders of magnitude slower than oxygen diffusion, it is a mechanism responsible for creep, and thus of interest where high temperature mechanical properties and phase stability of the material are of concern.

In this work we investigate both oxygen and cation diffusion in yttria-stabilized zirconia using a kinetic Monte Carlo (kMC) computer simulation procedure [3-5]. We discuss the effects on oxygen diffusivity of Y ion concentration and temperature, and also describe speculative simulations on the effects, on oxygen diffusivity, of the presence of oxygen vacancies in nearest neighbor configurations with respect to Zr or Y barrier cations. We also discuss the dependence of cation diffusivity on Y_2O_3 concentration, oxygen vacancy concentration, and temperature, and compare the oxygen and cation diffusion results.

YSZ exists in a cubic fluorite structure in which the Zr and Y cations are located at sites on a face-centered cubic sublattice, while the oxygen ions are located in a simple cubic sublattice whose lattice constant is one-half that of the cation sublattice. Both oxygen and cation diffusion take place via the hopping of ions to nearest-neighbor vacancies on their respective sublattices.

In view of the range of potential applications, considerable experimental and theoretical effort has gone into understanding the behavior of YSZ, and, in particular, oxygen diffusion in YSZ and related materials. Of interest here, computer simulations using a variety of techniques have been performed.

Schelling et al. [6] have investigated the cubic-to-tetragonal phase transition using molecular dynamics simulation, and have correctly predicted the experimentally-observed stabilization due to yttrium doping of ZrO_2. Similar work has been carried out by Fabris et al. [7]. Fevre et al. have investigated the thermal conductivity of YSZ using both Monte Carlo [8] and molecular dynamics [9] techniques. Krishnamurthy et al. have performed kinetic Monte Carlo (kMC) simulations of oxygen diffusion in YSZ [10] and similar compounds [11], and have produced diffusivities in reasonable agreement with experiment.

A number of other computational studies of oxygen diffusion in YSZ have been performed, including molecular dynamics studies by Kahn et al. [12], Okazaki et al. [13], Perumal et al. [14] and Shimojo et al. [15].

Cation diffusion in YSZ is less widely studied, but a number of studies do exist. Kilo et al. [16] analyze Zr diffusion from creep data, dislocation loop shrinkage data, and Zr tracer diffusion data to identify the defect(s) responsible for cation diffusion. They identify diffusion involving single cation vacancies as the most likely mechanism. However, measurements of activation enthalpies remain problematic; the relative ordering of the enthalpies for Zr and Y diffusion are not consistent among various studies.

Kilo et al. [17] perform simulations of cation diffusion in YSZ (as well as doped lanthanum gallates), using NPT molecular dynamics and a Buckingham-plus-Coulomb potential. They consider the hopping of cations via vacancy sites, introduced in the form of Schottky defects, at a mole fraction of 0.004, for yttria mole fractions of 0.11, 0.19 and 0.31. They find that the diffusion coefficients for Y and Zr are significantly different, with Y diffusion 3-5 times faster than Zr diffusion. They report calculated enthalpies, for YSZ having ytrria mole fraction 0.11, of 4.8eV (Y) and 4.7eV (Zr). These differ in both magnitude and ordering from experimental results from Kilo et al. [16], who report 4.6eV (Y) and 4.2eV (Zr). The molecular dynamics results also show that cation diffusivities are independent of Y_2O_3

concentration, or slightly increasing with increasing Y_2O_3 concentration, in contrast with experiment, which shows cation diffusivity to decrease with increasing Y_2O_3 concentration.

KINETIC MONTE CARLO METHOD

Kinetic Monte Carlo simulation differs from typical Metropolis Monte Carlo simulation in that it is explicitly aimed at the simulation of the dynamical evolution of a system at the atomic level. As such, it complements the more widely-used molecular dynamics (MD) method. While MD can produce a detailed trajectory for each particle in the simulation, accurately representing atomic vibrations in such simulations requires that the numerical integration of the equations of motion be carried out using a time step on the order of femtoseconds.

This restriction means that MD can be very inefficient when used to study so-called "infrequent events," notably the diffusive hopping of atoms among vacancy sites of interest here. In an MD simulation of a system containing only a small fraction of vacancies, a computational cell of reasonable size will experience a relatively small number of diffusive hops, with most of the computational resources spent computing the trajectories of atoms between the infrequent hops. By contrast, the kinetic Monte Carlo method allows one to concentrate on the events of interest, and to effectively consider only the average behavior of the system between such events, while giving up information on the detailed trajectories of all atoms in the simulation.

Briefly, the kMC process involves the creation of a catalog of all possible events (in this case, diffusive hops) accessible to the system, along with the corresponding event probabilities. One of the hopping events is chosen stochastically and executed. The event catalog is modified to reflect the new position of the hopping vacancy and the corresponding new events accessible to the system, and new event probabilities are computed. Finally, the simulation clock is advanced stochastically. The process is repeated until a sufficient number of events have occurred to generate statistically useful information on the properties of interest.

In more detail, it is known that a diffusive hop typically takes place on a time scale much slower that the typical period of atomic vibration, so that the system effectively loses any memory of the details of the hop, i.e. which of the vacancy's neighbor atoms was involved in the most recent hop. Each such hop may therefore be considered to be an independent event. The probability per unit time that a vacancy will undergo a hop is constant, and the survival probability is given by a decreasing exponential. It can be shown that the probability distribution of the time of first escape is given by $p(t) = k_{tot} \exp(-k_{tot}t)$, and the average time of first escape τ is given by

$$\tau = \int_0^\infty tp(t)dt = 1/k_{tot}$$ .Because all hopping events are independent, the effective total rate constant is just the sum of rate constants for all possible paths, with each rate constant determined by the height of the migration energy barrier in the direction of the hop:

$$k_{tot} = \sum_B k_{AB}$$

When the migration barrier energies are known, the hopping rates may be computed from $\nu_{AB} = \nu^0 \exp(-E_{AB}/k_BT)$ in which ν_{AB} and E_{AB} are the hopping rate and migration barrier energy for a hop between oxygen or cation sites A and B respectively, and ν^0 is the frequency factor. ν^0 is typically assigned a value between 10^{12} and 10^{13} for these materials; given that the measured diffusivities (both oxygen and cation) from different experimenters can differ substantially, we assume a value of 10^{13}

with the understanding that the values of the diffusivities presented here involve considerable uncertainty. For each possible hop, the hopping probability can be computed from the hopping rate, with $P_{AB} = v_{AB}/\Gamma$, where Γ is the sum of hopping rates for all possible hops in the computational cell. A catalog of all possible hops, and the corresponding hopping rates and probabilities, is created.

During the kMC process, one of the possible events (that is, a hop defined by the hopping ion and the target vacancy site) is chosen probabilistically from the catalog and executed. Hopping rates for all possible hops involving the new vacancy location are computed and added to the catalog, while rates involving the vacancy's previous location are deleted, and the sum of the hopping probabilities is updated. Finally, the simulation clock is advanced by a stochastically-chosen time step $\Delta t = -\ln(R)/\Gamma$ where R is a random number greater than zero and less than or equal to unity.

When the simulation has run long enough to accumulate statistically useful information, the mean square displacement, averaged over all vacancies, is computed. The vacancy diffusivity D_v is obtained from the Einstein relation $\langle R^2 \rangle = 6D_v t$, and the ionic diffusivity D_i is obtained by balancing the number of vacancy and ionic hops:

$$D_i = \frac{C_v}{1-C_v} D_v$$

where C_V is the concentration of vacancies on the oxygen sublattice.

The most energetically favorable hopping paths are in the [100] direction (oxygen), and [110] direction (cation). In the case of oxygen diffusion, this has been confirmed by ab initio calculations [10], while for cation diffusion, the direction is consistent with the MD simulations of Kilo [17].

As an initial test, we show results for the mean square vacancy displacement as a function of time (represented by the number of Monte Carlo steps) in Figure 1 for both oxygen and cation diffusion. Because the kMC process incorporates a random walk, the mean square displacement is expected to be linear in time (and in the number of diffusive steps), and this behavior is evident in the results.

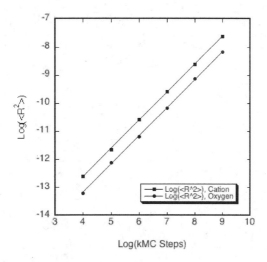

Figure 1. Mean square displacement versus kMC diffusive steps.

OXYGEN DIFFUSION

We have performed kMC simulations of oxygen diffusivity in YSZ for yttria mole fractions ranging from 0.01 to 0.25, and temperatures ranging 1500K to 2750K. The energetically-favored hopping path is in the [100] direction on the oxygen sublattice with the hopping oxygen ion passing between two barrier cations, as shown in Figure 2.

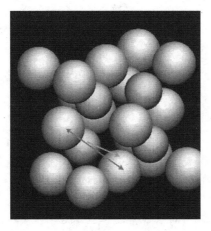

Figure 2. Oxygen ion hopping path (blue arrow). The two barrier cations are indicated by the red arrow.

Barrier energies were computed with the Abinit plane wave pseudopotential density functional code [23], using a 2x2x2 periodic supercell of fixed lattice constant, in which the atomic positions were fully relaxed. The barrier energies for all possible two-atom barriers are shown in Table 1, along with barrier energies for earlier kMC simulations computed using Car-Parinello molecular dynamics.

Table 1. Oxygen diffusion barrier energies, eV.

Authors	Method	Zr-Zr Barrier	Zr-Y Barrier	Y-Y Barrier
Krishnamurthy [10]	Car-Parinello MD	0.58	1.29	1.86
Krishnamurthy [11]	Car-Parinello MD	0.473	1.314	2.017
This work	DFT	0.706	1.214	1.941

MD calculations have yielded activation energies of 0.37 eV [12] and 0.2-0.8 eV [24]. A tracer diffusion study finds a value of 0.44 eV [25], while results from bulk conductivity and ac impedance spectroscopy give values of 0.79-1.12 eV [26-28].

Diffusivities over the above temperature range for different Y^{3+} concentrations are shown in Figure 3. In all cases, at low concentrations the diffusivity increases with concentration, reaches a maximum at about 0.08-0.15 mole fraction Y_2O_3, and decreases at higher concentrations. This behavior is similar to that observed by Krishnamurthy using a similar kMC method [10], and qualitatively consistent with experimental results [25]. In addition, the absolute values of diffusivity in this work are in reasonable agreement with experiment. Finally, the fact that the maximum occurs at higher Y concentrations at higher temperatures is consistent with observation.

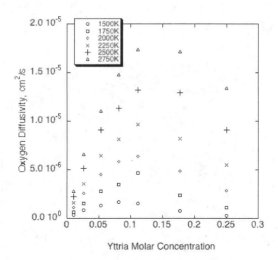

Figure 3. Oxygen diffusivity versus yttria concentration.

Arrhenius plots of the same data are shown in Figure 4, for a variety of Y_2O_3 concentrations. The slopes of the plots of our results are in reasonable agreement with experiment [13, 15, 24, 25], and with other simulations [12]. It can also be seen that the activation energies obtained from our data increase with increasing Y concentration, consistent with the inclusion of a greater number of higher-energy Zr-Y and Y-Y barriers.

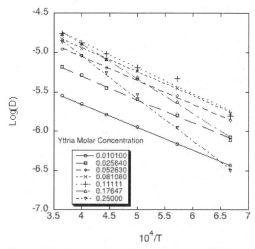

Figure 4. Temperature dependence of oxygen diffusivity.

CLUSTERING

There is evidence for the existence of various defect clusters in YSZ; at high Y^{3+} concentrations, complex defect configurations that may be considered as non-stoichiometric regions may be observed, but these are beyond the scope of the present work. At low Y^{3+} concentrations, the clusters of interest are simpler, for example, an oxygen vacancy lying on the oxygen sublattice in a nearest-neighbor or next-nearest-neighbor configuration with respect to a Zr or Y ion. If such an ion is a barrier ion, the value of the energy barrier may be affected by the presence of the neighboring vacancy.

We have performed density functional calculations to determine whether the presence of an oxygen vacancy in a nearest-neighbor position with respect to a Zr or Y barrier cation modifies the energy barrier to a significant degree.We have compared the energetics of a computational cell in which an oxygen vacancy exists as a nearest neighbor of a barrier cation (referred to as the "clustered" configuration) with those of a cell in which the oxygen vacancy is relatively far from the ion.

It should be noted that the 2x2x2 computational cell does not allow the vacancy to be very far from the ion (or its periodic image), so that the numerical value of the difference in energies may change somewhat when computed using a larger cell. However, the qualitative ordering of the energies is likely correct. We have examined four configurations, as shown in Table 2: oxygen vacancies adjacent to Zr barrier cations in Zr-Zr and Zr-Y barrier pairs, and adjacent to Y cations in Zr-Y and Y-Y pairs. Barrier energies for cases where oxygen vacancies are adjacent to both barrier cations are currently being computed, but these configurations are relatively rare. The barrier energies for the clustered configurations are reduced by six to fifty percent.

Table 2. Effects of neighboring oxygen vacancies on barrier energies.

Barrier Configuration	Barrier Energy, eV
Zr-Zr, no vacancy	0.706
Zr-Zr, nn vacancy	0.479
Zr-Y, no vacancy	1.314
Zr-Y, Zr-nn vacancy	0.593
Zr-Y, Y-nn vacancy	1.14
Y-Y, no vacancy	1.941
Y-Y. Y-nn vacancy	1.64

These results have been used in kMC runs in which the probabilities of the unclustered and clustered configurations are chosen consistent with the composition. Diffusivities for T=2250K are shown in Figure 5, for clustered and unclustered configurations, as a function of yttria concentration. The diffusivities for the clustered configurations are larger than those for the unclustered configurations in each case, except for the $C_{yttria} = 0.01$ results, where the diffusivities are approximately equal. The magnitudes of the differences range from approximately zero to eighteen percent (ignoring the unclustered value at $C_{yttria} = 0.18$, which appears to be an outlier), with the larger differences occurring at larger yttria concentrations. The increases in diffusivity in the clustered cases is consistent with the reduction of the corresponding barrier energies.

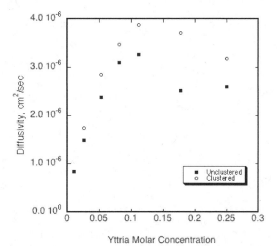

Figure 5. Oxygen diffusivity, T=2250K, effects of barrier reduction due to the presence of neighboring oxygen vacancies.

CATION DIFFUSION

Similar kMC simulations have been carried out for cation diffusion. The nearest-neighbor cation hop is in the [110] direction, through a barrier consisting of two oxygen ions, as shown in Figure 6. The DFT energies were computed with the position of the hopping ion held fixed at the presumed saddle point while all other atoms were allowed to relax. The resulting barrier energies, when incorporated into kMC calculations, produce cation diffusivities that are considerably too small. There is some evidence that the actual saddle point is not the one suggested by symmetry alone, and we are currently performing reaction path DFT calculations to investigate the issue.

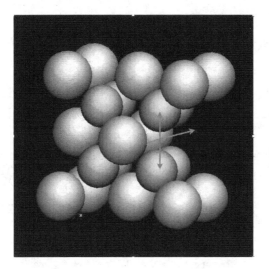

Figure 6. Cation hopping path (blue arrow). The two barrier oxygen ions are indicated by the red arrow.

In view of this difficulty, we have made use of two sets of barrier energies. One set (Set 1) includes barrier energies of 3.7eV (Y) and 3.62eV (Zr), obtained by fitting to experimental results from Kilo [16], with the energies in the same proportion as MD energies from MD simulation[17], so as to give results consistent with experiment at $C_{yttria}=0.1$, $C_{cation\ vacany}=0.004$, and T=2000K. The other set (Set 2) includes energies of 4.8eV (Y) and 4.2eV (Zr), taken from experimental results of Kilo [16]. Both of these sets of values, along with other energy and enthalpy results from experiment and simulation, are shown in Table 3. It should be noted that there is considerable variation, both in the values, and in the ordering, in the results from different researchers.

Table 3. Cation migration enthalpies and barrier energies form experiment, simulation and theory.

Authors	Temperature Range, C	Energies, eV
Solmon [15]	1300-1700	4.8-4.95
Gomez-Garcia [16]	Above 1500	5.5-6.0
Chien and Heuer [17]	1100-1300	5.3
Dimos and Kohlstedt [18]	1400-1600	5.85
Kilo [13]	1125-1460	4.4-4.8
Mackrodt [18]		2-7
This work, Set 1 (fitted)		3.7 (Y), 3.62 (Zr)
This work, DFT		6.55 (Y), 6.16 (Zr)
This work, Set 2 (expt)		4.8 (Y), 4.2 (Zr)

As previously mentioned, earlier studies of oxygen diffusion in YSZ, using similar methodology [10,11] found that the behavior of the diffusivity as a function of Y concentration was the result of a competition between the increased oxygen vacancy concentration that corresponds to an increase in Y concentration, and the increase in higher-energy hopping energy barriers due to the increased presence of Zr-Y and Y-Y barrier pairs.

The situation for cation diffusivity is somewhat different. The oxygen vacancy concentration is again tied to the concentrations of Y^{3+} cations as well as cation vacancies; in both cases, substituting a Y^{3+} cation or a cation vacancy for a Zr^{4+} cation requires additional oxygen vacancies to be created. As the cation vacancy concentration increases, there are two effects on cation diffusivity. First, the diffusivity will increase as the number of available vacancy sites available for hops increases. Second, the resulting increase in oxygen vacancies means the there is a larger number of oxygen barrier pairs in which one or both oxygen ions are replaced with vacancies. Initial DFT calculations suggest that even a single vacancy in the barrier pair lowers the barrier energy substantially, which will increase the diffusivity. However, the cation vacancy concentration is typically much smaller than the Y^{3+} concentration, so that the effect of increasing the cation vacancy concentration on vacancy population in the barriers will be small, and we neglect the effect in these simulations; all barriers are assumed to be fully populated, without vacancies. The result is that the dependence of the diffusivity on cation vacancy concentration is dominated by the increase in available hopping sites, with the result that the diffusivity should increase with cation vacancy concentration. This behavior is exhibited by our KMC simulations, as shown in Figure 7, where the cation diffusivity is seen to increase linearly with cation vacancy concentration.

The dependence of cation diffusivity on Y^{3+} concentration is more complicated, and results of other researchers are ambiguous. As the Y^{3+} concentration increases, the concentration of oxygen vacancies increases, resulting in more vacancy-containing barrier pairs, as described above. However, because the Y^{3+} concentration is typically much larger than the cation vacancy concentration, the effect in this case is not necessarily negligible. If the barrier energies are lowered significantly by the presence of a vacancy at one of the barrier sites, the effect on diffusivity may be significant. We are performing DFT calculations of vacancy-containing barriers with the aim of resolving the issue. Given the difference in ionic radii between oxygen and the two cations, it is expected that even a single-vacancy barrier will yield a significantly smaller barrier energy, so that cation diffusivity at large Y^{3+} concentrations may be enhanced. The calculation of accurate energies in such a reduced-symmetry situation will likely require the same reaction path approach discussed previously. Our current simulations to not incorporate the effect.

In addition, increasing the Y^{3+} concentration also increases the ratio of Y to Zr cation hops, and the effect of this on the diffusivity is not clear due to the uncertainty in the barrier energies, and their ordering. Regardless, it is likely that the difference in energy barriers for Zr and Y hops is relatively small, so that the direct effect of Y^{3+} concentration on diffusivity is probably small. Depending on the relative sizes of the barriers, increasing the Y^{3+} concentration may have either the same, or the opposite, effect as increasing the cation vacancy concentration, so that either a weak increase or a weak decrease in diffusivity with increasing Y^{3+} concentration is plausible. Because we have yet to compute barrier energies for vacancy-containing barriers,

KMC simulations have been performed for a range of compositions, with a Y^{3+} mole fraction of 0.1. A range of cation mole fractions from 0.001 to 0.01 is considered, and the concentration of oxygen vacancies is adjusted in each case to guarantee cell neutrality.

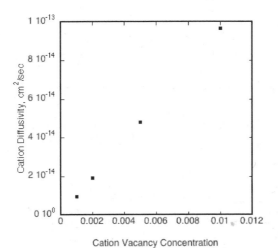

Figure 7 Cation diffusivity versus cation vacancy mole fraction, T=2250K.

The temperature dependence of cation diffusivity is shown in Figure 8, along with experimental results. Two results from the current work are shown, using the two energy sets shown in Table 3.

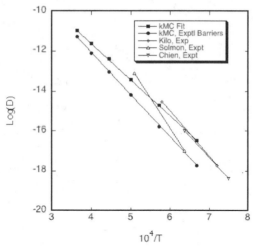

Figure 8. Temperature dependence of cation diffusivity.

Because the diffusive energy barriers in Set 1 were chosen to duplicate one experimental result at 2000C, the agreement between this work and experiment is not unexpected. The slope of the line from our KMC calculations is less than that of the experimental results shown, although it differs from the results of Chien et al. by less than ten percent. Our results using Set 2 energies are smaller than experimentally observed diffusivities, though the slope of the line is consistent with most other results. It may be the case that when the effect of vacancies in the barriers is understood, these energies may yield results in better agreement with experiment.

Other barrier energies show a slope more consistent with other experimental results, although, as described above, the absolute values of the diffusivity are smaller. In any event, the reasonable agreement between simulation and experiment of the slopes of the ln(D) versus 1/T suggests that the KMC simulations capture the fundamentals of the diffusion process.

CONCLUSIONS
We have performed kinetic Monte Carlo computer simulations of oxygen and cation diffusion in yttria stabilized zirconia. Cation diffusivities computed here are several orders of magnitude lower than oxygen diffusivity in the same materials, qualitatively consistent with experimental observation, and with molecular dynamics results. Oxygen diffusivities computed using barrier energies from DFT calculations are consistent in magnitude with results from experiment and from other simulations. The temperature dependence of the diffusivity is also consistent with experiment, and activation energies show the expected increase with increasing yttria concentration.

The presence of an oxygen vacancy is a nearest neighbor to one of the two cation barrier ions reduces the barrier energy and tends to increase the oxygen diffusivity. Additional configurations are being investigated, as are the effects of larger complexes of these clusters.

Our initial DFT cation hopping barrier energies produce diffusivities that are too large. In order to produce cation diffusivities consistent with experiment, we have used two sets of hopping barrier energies that are within the range of experimental values, one of which includes energies that are towards the smaller end of the range of experimental numbers. The temperature dependence of the kMC results is in reasonable agreement with experiment, as is the qualitative dependence on cation vacancy concentration.

REFERENCES

[1] P. Aldebert and J. P. Traverse, J. Am Ceram. Soc. 68 [1], 34-40 (1985).
[2] R. E. W. Casselton, Phys. Status Solidi A, 2, 571-585 (1970).
[3] W. M. Young and E. W. Elcock, Proc. of the Phys. Soc. 89, 735 (1966) .
[4] A. B. Bortz and M. H. Kalos and J. L. Lebowitz, J. of Comput. Physics 17, 10 (1975).
[5] A. F. Voter, Kinetic Monte Carlo, in Radiation Effects in Solids, Proceedings of the NATO Advanced Study Institute on Radiation Effects in Solids, K. E. Sickafus, E. A. Kotomin, B. P. Uberuaga, eds., 1-23, Springer, Dordrecht, The Netherlands, 2007.
[6] P. K. Schelling, S. R. Phillpot and D Wolf, J. Am Ceram. Soc. 84 [7], 1609-1619 (2001).
[7] S. Fabris, A. T. Paxton and M. W. Finnis, Phys. Rev. B 63, 094101 (2001).
[8] M Fevre, A. Finel and R. Caudron, Phys. Rev. B 72, 104117 (2005).
[9] M Fevre, A. Finel, R. Caudron and R. Mevrel, Phys. Rev. B 72, 104118 (2005).
[10] R. Krishnamurthy, Y.-G. Yoon, D. J. Srolovitz and R. Car, J. Am. Ceram. Soc. 87 [10],1821-1830 (2004).
[11] R. Krishnamurthy, D. J. Srolovitz, K. N. Kudin and R. Car, J. Am. Ceram. Soc. 88 [8],2143-2151 (2005).
[12] M. S. Kahn, M. S. Islam and D. R. Bates, J. Mater. Chem. 8 [10], 2299-2307 1998.
[13] H. Okazaki, H. Suzuki and K. Ihata, Phys. Let. A 188, 291-295 (1994).
[14] T. P. Perumal, V. Sridhar, K. P. N. Murthy, K. S. Easwarakumar and S. Ramasamy, Comp. Mat. Sci. 38, 865-872 (2007).
[15] F. Shimojo, T. Okabe, F. Tachibana, M. Kobayashi and H. Okazaki, J. Phys. Soc. Jpn 61, 2848-2857 (1992), and F. Shimojo and H. Okazaki, J. Phys. Soc. Jpn 61, 4106-4118 (1992).
[16] M. Kilo, G. Borchardt, C. Lesage, O. Kaitsov, S. Weber and S. Scherer, J. Eur. Ceram. Soc., Faraday Trans. 5, 2069 (2000).
[17] M. Kilo, M. A. Taylor, C. Argirusis, G. Borchardt, R. A. Jackson, O Schulz, M. Martin and M. Weller, Solid State Ionics 175, 823-827 (2004).
[18] H. Solmon, C. Monty, M. Filial, G. Petot-Ervas and C. Petot, Solid State Phenom. 41, 103 (1995).
[19] D. Gomez-Garcia, J. Martines-Fernandez, A. Dominguez-Rodriguez and J. Castaing, J. Am. Ceram. Soc. 80, 1668-1672 (1997).
[20] F. R. Chien and A. H. Heuer, Philos. Mag. A 73, 681-697 (1996).
[21] D. Dimos and D. L. Kohlstedt, J. Am. Ceram. Soc. 70, 277 (1987).
[22] W. C. Mackrodt and P. M. Woodrow, J. Am. Ceram. Soc. 68, 277 (1986).
[23] X. Gonze, B. Amadon, P.-M. Anglade, J.-M. Beuken, F. Bottin, P. Boulanger, F. Bruneval, D. Caliste, R. Caracas, M. Cote, T. Deutsch, L. Genovese, Ph. Ghosez, M. Giantomassi, S. Goedecker, D. R. Hamann, P. Hermet, F. Jollet, G. Jomard, S. Leroux, M. Mancini, S. Mazevet, M. J. T. Oliveira, G. Onida, Y. Pouillon, T. Rangel, G.-M. Rignanese, D. Sangalli, R. Shaltaf, M. Torrent, M. J. Verstraete, G. Zerah, J. W. Zwanziger, Computer Phys. Comm. 108, 2582 (2009).
[24] X. Li and B. Hafskjold, J. Phys.: Condens. Matter 7, 1255 (1995).
[25] Y. Oishi and K. Ando, Transport in Nonstoichiometric Compounds, G. Simovich and V. Stubican, eds., NATO ASI Ser. B 129, 1985.
[26] I. R. Gibson and J. T. S. Irvine, J. Mater. Chem. 6, 895 (1996).

[27] A. Orliukas, P. Bohac, K. Sasaki and L. J. Gauckler, Solid State Ionics 72, 35 (1994).
[28] I. R. Gibson, E. E. Lachowski, J. T. S. Irvine and G. P. Dransfield, Solid State Ionics 72, 265 (1994).

Advanced Sensor Technology

NANO-CALORIMETER PLATFORM FOR EXPLOSIVE SENSING

Kang, Seok-Won; Niedbalski, Nicholas; Lane, Mathew R.; Banerjee, Debjyoti
Mechanical Engineering Department, Texas A&M University,
College Station, TX, United States

ABSTRACT

The aim of this study is to develop a robust field deployable portable nano-calorimeter sensor for detection of explosive vapors, typically emanating from Improvised Explosives Devices. The microcantilever sensors are composed of two layers: *400 nm Au* film on a *600 nm Si_3N_4* substrate. The microcantilever bends in response to induced thermal stresses arising from temperature changes and the dissimilar Thermal Coefficients of Expansion (bimetallic actuator). The differences in bending response of the sensor arrays to adsorption and combustion reactions (catalyzed by the gold surface) are reported in this study. Different combustible materials such as alcohol or acetone were tested for detection by the sensor arrays. Numerical models were developed to predict the bending response of the microcantilevers in different environmental conditions. Joule-heating in the resistive heating element (Au) was coupled with the gaseous combustion at the heated surface to obtain the temperature profile and therefore the deflection of a microcantilever by calculating the thermo-mechanical stress-strain relationships. The sensitivity of the threshold current of the sensor that is used for the specific detection and identification of individual explosives samples - is predicted to depend on the reaction rates and the vapor pressure.

INTRODUCTION

Miniaturized microfabricated calorimeters have enormous potential in transduction of physical or chemical properties of very small length scales (i.e. nano-meter range) or quantities/ volumes (i.e. pico-liter range). Microcantilever based sensors have been widely used for chemical and biological detection applications. Other applications for the microcantilever sensors include pH sensing, pathogen sensing, and sensing by means of DNA hybridization [1]. There are two primary modes of operation of microcantilever sensors: static and dynamic modes [2-3]. In static mode, static response due to bending induced by differential surface stress is monitored. In dynamic mode the change in resonant frequency upon mass uptake are monitored. The adsorption/ desorption of molecules on the surface of microcantilevers causes the changes in surface stress (i.e. compressive or tensile stresses), which results in variations of harmonic or static responses.

Traditionally, in civilian and military operations, trained dogs are used to detect explosive materials. In military operations other trained animals/ insects (such as dolphins and bees) have also been used for explosives detection (in exotic environments such as in marine environments). To obviate practical issues concerning the usage of animals to detect explosives, automated procedures involving multiple types of electronic sensors ("electronic-nose" for air borne sensors and "electronic-tongue" for water based sensors) have been designed to replace the trained animals. Various electronic noses include the usage of fiber optics and beads, polymeric films, gold nanoclusters, surface acoustic waves (SAW), and micro-electrochemical systems (MEMS) [3]. In this study, micro-electrochemical systems were used as an electronic nose to detect explosives. This idea comes from the endeavor to mimic bomb-sniffing dogs, so we usually refer to the technology as electronic or artificial "noses". An electronic nose is typically composed of an electrical power system, a chemical sensing system, and a response detection system.

Sensor arrays based on this system have multiple advantages such as their extremely small size, fast response time, very low detection threshold, and multiple component analysis capability [1-5]. Microcantilever-based explosive gas sensing is usually performed by utilizing different strategies, which include: (1) detecting adsorption of molecules of interest on surface of the microcantilevers; (2)

recognizing molecules by the measurement of changes in bending response (or changes in surface shear stress); or (3) changes in resonance frequency (or changes in mass).

The sensor used in this study is based on a microfabricated array of bimorph microcantilever (ActivePen™ purchased from Nanoink Inc.). This product is fabricated by the thermal compression bonding at high temperature (~*300°C*) of *Au* film of *400 nm* deposited on Si wafer by evaporation at low pressure (*2 ~ 7 × 10⁻⁷ torr*) and *Si₃N₄* microcantilever of *600 nm* made by etching and metallization on oxidized Si wafer. The bi-morph cantilever array is heated individually by an array of gold microheaters that are microfabricated in-situ at the base of each cantilever. The microcantilever provides a structural response (i.e. bending motion) to the variations in thermal stress at the surface, caused by Joule heating (i.e., on supplying different amounts of electrical power to the micro-heaters).

In this study we report the design analysis of the performance of the nano-calorimeter platform by utilizing electrically pre-heated microcantilever arrays for absorption, catalytic oxidation or surface reaction with specific volatile and combustible gases. The deflection is experimentally measured using optical detection method by tracking the light spot reflected from the microcantilever surface. We additionally conducted a numerical analysis to verify the measurements.

NUMERICAL MODELING

Computational model development and simulations were performed using a commercial finite element analysis (FEA) tool (Ansys®). The computational model was used to perform a parametric study of the coupled electro – thermo - mechanical analyses of the microcantilever platform used in this study. Proper estimate for the temperature profile of the microcantilevers is a key factor in our nano-calorimeter platform for chemo-mechanical sensing of explosives. However, the thermal response to chemical reactions is not available in the FEA tool (Ansys®). Our chemo-mechanical model predicts the mechanical deflection for a change in bimorph temperature. So, we demonstrate the procedure of electro – thermo - structural modeling by coupling of a Computational Fluid Dynamics (CFD) tool (Fluent®) with an FEA tool (Ansys®), as shown schematically in Figure (1). The thermal data obtained from the chemical reaction model (using Fluent®) is mapped onto each Finite Element (FE) node, which serves as the initial condition for the structural dynamics simulation using Ansys®.

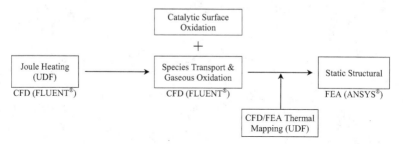

Figure 1. Schematic of a complete model of an electro-thermally actuated microcantilever

Computational Fluid Dynamics (CFD)

The volumetric Joule heat generation, q, by electric current through resistive heating element can be calculated from Ohm's Law as follows:

$$Q = I^2 R = I^2 \rho \frac{l}{A} \quad \rightarrow \quad q = \frac{Q}{V} = \left(\frac{I}{A}\right)^2 \rho \qquad (1)$$

where I is the applied current [A], ρ is the resistivity [Ωm], A is the cross-sectional area [m^2], and l is the length of resistive element [m]. Fluent® does not provide the solution for joule heating, so we implemented the user-defined function (UDF) code into the Fluent® case file. In our UDF code, the electrical conductivity value is defined as the diffusivity of the solid phase potential in the solid zones. Then we perform the thermal analysis for the catalytic oxidation based on species transport and gas phase as well as surface oxidation models. The Au catalysis used in this study enables ultra lean oxidation, which means it is available to detect the gas vapor of very low concentrations. It is also expected that the gas-phase combustion occurs in a "flameless" mode near the heating element [6].

Simulations were performed at a constant gas mole (or mass) fraction in a testing chamber. The vapor-liquid equilibrium (VLE) mole fraction is determined from an evaporation pressure at a room temperature ($298\ K$). In Fluent®, concentrations of reactants need to be specified on the basis of mass fractions. The 'Laminar finite-rate model' was selected in Fluent®, which is of the following form:

$$k = A_r T^{\beta_r} \exp\left(-\frac{E_r}{RT}\right)$$ (2)

where, A_r is the pre-exponential factor [s^{-1}], βr is the temperature exponent, E_r is the activation energy for the reaction [J/kmol], and R is the universal gas constant [J/kmol-K]. The complete oxidation reaction of VOC is highly exothermic and the global oxidation models of different gases (that were used for numerical simulations) are summarized in Table I.

Table I. Global one-step reaction models of Acetone and Isopropyl Alcohol

Gases	Combustion Model	Heat of Combustion
Acetone $(CH_3)_2CO$	$(CH_3)_2 CO + 4O_2 \rightarrow 3CO_2 + 3H_2O$	-1761 kJ/mol (-303.2×10^5 J/kg)
Isopropyl Alcohol C_3H_7OH	$2C_3H_7OH + 9O_2 \rightarrow 6CO_2 + 8H_2O$	-1907 kJ/mol (-317.3×10^5 J/kg)

The gas phase reaction scheme is taken from various literatures, which was modeled based on 1st order homogeneous reactive flow proportional to the volumetric concentrations of species. Basically, in the gas phase reactions of VOC, Hydrogen abstraction leads to the formation of CO2 or H2O as the results of deep (or complete) oxidation. Also, especially at low temperature, VOC is oxidized to the intermediate products (i.e. Acetone: CO/ IPA: CO, C_3H_6 or C_3H_6O). The multiple-step combustion model as shown in Table II gives a more optimized value as opposed to the "global" one-step reaction model.

In general, the catalytic oxidation by surface reaction depends on the surface coverage of explosives. The surface reaction rate for VOC over the catalytic surface deposited with Au catalyst is calculated by using Equation (3), proposed by Hayes and Kolaczkowski [10].

$$R_s = \eta A_s \exp\left(-\frac{E_s}{RT}\right)Y_{VOC,s}$$ (3)

The effectiveness factor in Equation (3), η, addresses the diffusion effect of reactants in the catalyst, which can be calculated by Equation (4) and (5), as derived as [10].

$$\eta = \frac{\tanh(\phi)}{\phi}$$ (4)

$$\phi = L_c \sqrt{\frac{k_s}{D_{eff}}} \tag{5}$$

where ϕ is the Thiele modulus, L_c is the thickness of the catalyst, k_s is the rate constant based on catalyst surface area $[kg/m^2 s]$, and D_{eff} is the effective diffusion coefficient $[m^2/s]$.

Table II. Chemical Kinetic Parameters for Gas Phase Reaction of Acetone and Isopropyl Alcohol

Reaction	A_r	β_r	E_r
$2C_3H_6O + 5O_2 \rightarrow 6CO + 6H_2O$ [7-8]	1.9×10^{11}	-1.0	2.09×10^8
$2H_2 + O_2 \rightarrow 2H_2O$ [7-8]	2.37×10^{-3}	-0.5	8.79×10^7
$2CO + O_2 \rightarrow 2CO_2$ [7-8]	3.55×10^5	-1.5	8.79×10^7
$CO + H_2O \rightarrow CO_2 + H_2$ [7-8]	1.2×10^8 3.73×10^9	-1.0 -1.0	1.74×10^8 2.06×10^8
$C_3H_7OH \rightarrow C_3H_6 + H_2O$ [9]	8.32×10^7	0.0	8.11×10^7
$C_3H_7OH + 1/2O_2 \rightarrow C_3H_6O + H_2O$ [9]	4.58×10^7	0.0	7.16×10^7
$C_3H_6 + 9/2O_2 \rightarrow 3CO_2 + 3H_2O$ [9]	6.75×10^9	0.0	9.57×10^7
$C_3H_6O + 4O_2 \rightarrow 3CO_2 + 3H_2O$ [9]	1.39×10^{20}	0.0	2.103×10^8

In catalytic combustion reactions at elevated temperatures, the overall rate of reaction becomes limited by diffusion in the catalyst [10]. The effective diffusion coefficient is theoretically determined as follows:

$$D_{eff} = \left(\frac{1}{D_{ab}} + \frac{1}{D_{kn}} \right)^{-1} \tag{7}$$

where D_{ab} is the bulk diffusion coefficient and D_{kn} the Knudsen diffusion coefficient respectively, which are given by [11]

$$D_{ab} = \frac{0.00143 T^{1.75} \left[\frac{M_1 + M_2}{M_1 M_2} \right]^{0.5}}{p \left[V_1^{1/3} + V_2^{1/3} \right]^2} \tag{8}$$

$$D_{kn} = \frac{4}{3} d_p \sqrt{\frac{RT}{2\pi M}} \tag{9}$$

where M_1 and M_2 are the molecular weight of component 1 and 2 $[kg/mol]$, V_1 and V_2 are the molar volume of component 1 and 2 with the pressure p and temperature T $[m^3/mol]$, and d_p is average micropore diameter of adsorbent $[m]$. The surface reaction of VOC and those chemical kinetics parameter values are listed in Table III. Numerical simulations are performed based on the 3-D, laminar, species transport, gas phase as well as surface reaction, and steady-state simulation techniques.

Hexagonal and gradient meshing techniques are used. Since the testing chamber is closed, evaporation occurs at a state of dynamic equilibrium. Figure (2) shows the solid model generated in Gambit® software for thermal analysis with Fluent®.

Table III. Chemical Kinetic Parameters for Surface Reaction of Acetone and Isopropyl Alcohol

Reaction	A_s	E_s
$C_3H_6O + 4O_2 \rightarrow 3CO_2 + 3H_2O$ [12]	7.63×10^{13}	2.38×10^7
$C_3H_7OH + 0.5O_2 \rightarrow C_3H_6O + H_2O$ [13]	1.8×10^7	1.6×10^6

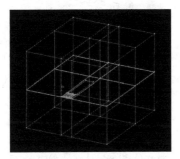

Figure 2. Geometry of the control volume for simulation.

Finite Element Analysis (FEA)

The Computational Fluid Dynamics (CFD) tool provides a powerful and flexible numerical framework for modeling fluid flow and performing associated convection heat transfer calculations, but does not have built-in advanced solid mechanics analysis capabilities for performing thermo-mechanical stress analysis. On the other hand, the Finite Element Analysis (FEA) tool provides the advanced solid mechanics analysis capabilities. To calculate the mechanical deflection by the thermal stress at the surface of microcantilevers, we implement UDF code for CFD/FEA thermal mapping into Fluent® calculation. The meshing and scaling of the models should be consistent in both Fluent® and Ansys®, as shown in Figure (3).

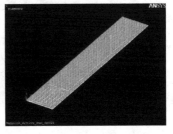

Figure 3. Solid Model of Microcantilever in (LEFT) Gambit® and (RIGHT) ANSYS®

After mapping of the solid temperature data from Fluent®, we calculate the mechanical deflection using FEM software Ansys® Multiphysics v12. The temperature data should be imported before determining the solver type and boundary conditions. The simulations were performed on three-dimensional FE models of the cantilevers that re assumed to be under linear and static conditions. The FE models were meshed by SOLID226 elements.

EXPERIMENTAL MEASUREMENT

The experimental setup consists of an air-tight acrylic chamber, a platform to support and control the movement of the laser, a platform for the microcantilever beam, and a piece of paper to mark the location of the reflected laser beam spot. The experimental apparatus is placed inside an environmental control chamber, which is constructed from rectangular acrylic walls with a hinged door made from ½" thick acrylic sheets as shown in Figure (4). In order to render the chamber airtight, silicone was used to seal the edges inside the box and a tape insert was used to seal the edges along the exterior of the box. Weather-stripping was used as a sealant between the door and the front wall of the chamber. Inside the chamber, a low-power laser (1 *mW*, 635 *nm*) was affixed to a semi-automated stage with 4 axes of motion (assembled from Newport components). The Newport stage system (Figure 4) supports the laser and can be actuated remotely for laser beam alignment with the cantilever axes in the nano-calorimeter apparatus. In addition to altering the position of the laser, the remote control can also be used to rotate the cantilever array within the *xy*-plane. Hence, laser alignment and cantilever positioning can be accomplished without disturbing the chamber environment.

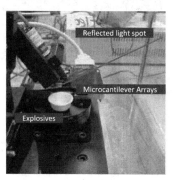

Reflected light spot

Microcantilever Arrays

Explosives

Figure 4. Experimental Apparatus for Explosive Detection

Each experiment was performed in two separate steps: a control experiment (baseline) was performed in ambient atmospheric conditions, and a second experiment was performed in the presence of the explosive vapor at equilibrium vapor concentration. The deflection response of the microcantilever (as a function of actuation current) in uncontaminated air environment was compared to that of air saturated with the explosives vapor. As shown in Figure (4), the laser beam incidenton the microcantilever surface is reflected by the gold coating on to the screen (paper). The actuation current (for heating the microheaters and therefore for actuating the microcantilever beam) is increased from *0* to *20 mA* at *2 mA* intervals. It was observed that at actuation current values exceeding 30mA - the bending response diverged from the control experiments. For each *2 mA* increment, the resulting deflection of the microcantilever beam is tracked by measuring the location of the laser beam spot reflected on the screen (paper) attached to the chamber wall. The deflection is expected to be proportional to the change in the location of the reflected laser spot, whichis measured by noting the difference in vertical location of the laser spot centroid from a reference position (in this case, the laser beam position at 0 mA actuating current). The resistance of the gold filament is also recorded at each current increment. Once the data is collected for the control experiment, the liquid explosive is poured into a small bowl and placed inside the chamber. The liquid remains in the chamber for approximately thirty minutes to ensure saturation with explosive vapor. Subsequently, the experiment is repeated in the presence of vapor samples. The results are then recorded and compared to the results obtained from the control experiments.

RESULTS AND DISCUSSION

Oxidation of Volatile Organic Compounds (VOC) with air is numerically studied at specific initial concentrations in a finite control volume. The chemical kinetics expressed in an Arrhenius form as written in Equation (2) are used to model the temperature dependence of reaction rate and activation energy for oxidation. The higher surface area to volume ratio at the nano-scale is expected to expedite the kinetics of the area-limited catalytic reactions, which means the chemical reactions only occur on the catalyst surface provided by the gold coatings on the microcantilevers [6]. The catalytic reaction on the surface of the microcantilevers depends on the core temperature of the heating element. Figure (5) shows the surface temperature range of electrically pre-heated microcantilevers in air. These results can be easily obtained from the electro-thermo coupling simulation in Ansys$^{®}$.

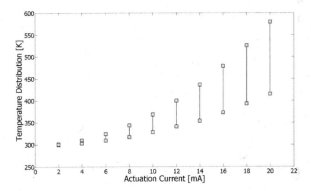

Figure 5. Temperature range of microcantilever heated in air.

However, to estimate the variations of surface temperature after oxidation on the pre-heated catalyst surface, we should simultaneously solve the mass and heat transfer problems with chemical reaction and resistive heating together. Since the mapping of temperature data from Ansys$^{®}$ to Fluent$^{®}$ is not supported, so we implement the UDF code for resistive heating into the chemical reaction model. To verify the UDF code, we compare the temperature profile calculated by UDF for Ohmic-heating in Fluent$^{®}$ with the results of electro-thermo multiphysics in Ansys$^{®}$.

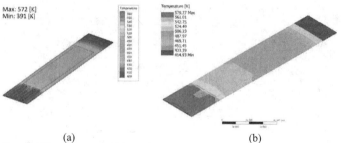

(a)   (b)

Figure 6. Temperature profile of bimorph microcantilever due to ohmic-heating
(a) UDF in FLUENT® (b) Multiphysics in ANSYS®

Figure (6) represents the comparison of numerical simulation when we use the two different simulation tools. We use the same values for geometrical and material properties of microcantilevers as summarized in Table IV. Also, the temperature dependence of specific heat (Figure (7)) and enthalpy values (Table V) of VOC are considered to obtain more accurate estimate for the results.

Table IV. Material properties used in this study

Property	Silicon Nitride (Si_3N_4)	Gold (Au)
Thermal Conductivity (k) [W/mK]	*1.7* [14]	*150* [15]
Thermal Expansion Coefficient (α) [K^{-1}]	*0.3E-06* [14]	*14.6E-06* [16]
Elastic Properties E: Young's Modulus [GPa] / v: Poisson's Ratio	*224.6* [14] / *0.253* [17]	*74.5 / 0.35* [18]
Electrical Resistivity (R) [Ωm]	*1E+10* [19]	*2.214E-08* [20]

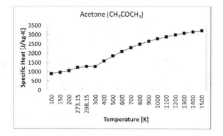
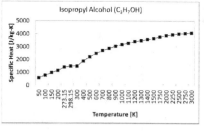

Figure 7. Temperature dependence of specific heat (C_p) of acetone and isopropyl alcohol

Table V. Initial concentration based on evaporation pressure and enthalpy for phase change of VOC samples used in this study

	Acetone $(CH_3)_2CO$	Isopropyl Alcohol C_3H_7OH
VLE mole Fraction [5] (Evaporation Pressure)	0.2 [mole] (186 [mmHg])	0.04 [mole] (33 [mmHg])
Standard State Enthalpy [21]	-2.19×10^8 [J/kgmol]	-2.73×10^8 [J/kgmol]
Standard State Entropy [22]	200,400 [J/kgmol-K]	180,580 [J/kgmol-K]

Oxidation of VOC proceeds with the formation of oxidation products carbon dioxide (CO_2) and water vapor (H_2O) as summarized in Table I. The initial conditions for concentration and the enthalpy for phase change are listed in Table V. Figure (8) shows the concentration profiles for reaction products for catalytic oxidation of acetone and isopropyl alcohol over the microcantilevers for an actuation current of 20 mA. At the applied current of 20mA, as shown in Figure (6a) the maximum surface temperature is 572 [K] that is below the ignition temperature (e.g. Acetone: 738 [K]/ Isopropanol: 672 [K]) reported for macro-scale.

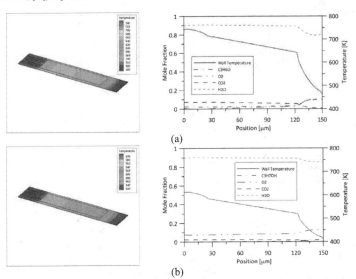

Figure 8. Surface temperature profile and coverage of species for nano-scale combustion reactions on the surface of the microcantilevers (a) Acetone (b) Isopropanol.

So, during the oxidation of acetone and isopropanol complete oxidation products such as CO_2 and H_2O are formed - as well as intermediate products such as CO are formed due to partial oxidation.

However, since the evaporation pressure of acetone is higher than that of isopropanol, it is observed that the oxidation was more active in the case of Acetone. Accordingly, the surface temperature increased by oxidation is more significant in Acetone. Figure (8) also shows the surface temperature of microcantilevers obtained from the simulations for the nano-scale combustion reactions. The temperature is increased by the presence of heat generation that is formed due to the oxidation reactions. Accordingly, the microcantilevers have a downward bending response, as shown Figure (9). Finally, the change in surface temperature due to combustion of VOC vapors contributes to the differences in deflections caused by the bimetallic effect.

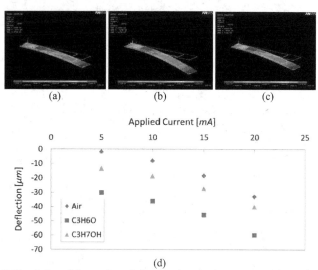

(a)   (b)   (c)

(d)

Figure 9. Simulation of the resultant deflection changes due to nano-scale combustion at 20 mA actuation current (a) Air (b) Acetone (c) Isopropyl Alcohol
(d) Comparison of deflection between air and explosives.

The nano-calorimeter was tested by performing experiments using Alcohol and Acetone as the sources for pure vapor. Figure (10) shows the results for the change in height of the reflected beam as a function of actuation current for four explosive materials. In the low current region, the microcantilever bending response in the presence of the combustible vapors is almost the same as the bending angles measured in air. As the electric current increases, the temperature of the bimorph microcantilever structure is increased causing additional bending of the beam and the incident laser ray is reflected from the microcantilever begins to deflect upwards causing the reflected light beam to move upwards. The presence of the vapor causes more vigorous oxidation on the surface of the microcantilever at elevated temperatures due to higher actuation current. The value of the actuation current at which the change in deflection deviates from the control experiments (performed in air) is called the threshold current. The values are used for uniquely detecting and indicating the specific explosives. For solid explosives, the threshold current can be estimated from the self-ignition temperature for a particular combustible vapor.

In addition, since Acetone or Isopropanol is highly volatile, their ignition temperatures as well as vapor pressures are key factors for predicting the threshold current. Their vapor pressures are summarized in Table V. The response to chemical reaction is well occurred at the lower temperature region than the self ignition temperature value due to those volatile property. The vapor pressure (186 mmHg) of Acetone is much higher than that (33 mmHg) of Isopropyl alcohol. So, the combustion reaction of acetone is activated at lower actuation currents. Additionally, the activation energy (301.1 × 106 [J/kgmol]) of isopropanol is higher than that (137.7 × 106 [J/kgmol]) of Acetone, which means the combustion reaction of Isopropyl Alcohol requires higher energy (or temperature). It is observed that the deflection characteristics in VOC show the same tendency (downward bending response) from the numerical and experimental results.

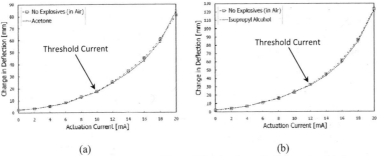

(a)   (b)

Figure 10. Experimental results of microcantilever deflection in explosive sensing (a) Acetone (Self Ignition Temperature = 738.15 [K]) (b) Isopropanol (Self Ignition Temperature = 672.15 [K])

SUMMARY AND CONCLUSION

In this study the static response of a microcantilever in the presence of explosive or combustible vapors were characterized experimentally and by performing numerical simulations. To explore the bending response, we experimentally measured the change in deflection of the microcantilevers caused by bimetallic effect - as a function of actuation current. Additionally, we performed the numerical analysis based on electro-thermo-mechanical coupling model by using UDF in Fluent® and Ansys®. The sensor sensitivity can be enhanced by specifically coating the high thermal conductivity materials onto the micro-cantilever surfaces (e.g. using Dip-Pen Nanolithography: DPN). Ultimately, this approach can be implemented into a portable detection platform or integrated instrument for remote monitoring and real-time detection of explosives.

REFERENCES

[1]H.P. Lang, C. Gerber, *Handbook of Nanophysics: Nanomedicine and Nanorobotics - Nanomechanical Sensors for Biochemistry and Medicine*, 1st Ed., CRC Press (2010).

[2]E. Comini, G. Faglia, and G. Sberveglieri, *Solid state gas sensing-Chap 9. Cantilever-based gas sensing*, 1st Ed., Springer, New York (2009).

[3]S. Singamaneni, M.C. LeMieux, H.P. Lang, C. Gerber, Y. Lam, S. Zauscher, P.G. Datskos, N.V. Lavrik, H. Jiang, R.R. Naik, T.J. Bunning, and V.V. Tsukruk, Bimaterial microcantilevers as a hybrid sensing platform," *Adv. Mater.*, **20**, 653-680 (2008).

[4]J. Yinon, Detection of explosives by electronic noses, *Analytical Chemistry*, March 1 (2003).

[5]I.C. Nelson, D. Banerjee, W.J. Rogers, and M.S. Mannan, Detection of explosives using heated micro-cantilever sensors, *Proc. Of SPIE*, **6223**, 62230O-1-8 (2006).

[6]J.A. Ahn, C. Eastwood, L. Sitzki, and P.D. Ronney, Gas-phase and catalytic combustion in heat-recirculating burners, *Proc. of the Combustion Institute*, **30**, 2463-2472 (2005).

[7]V.Ya. Basevich, A.A. Belyaev, and S.M. Frolov, Global kinetic mechanism for calculating turbulent reactive flows, *Chem. Phys. Reports*, **17**, 1747-1772 (1998).

[8]N.N. Gnesdilov, K.V. Dobrego, I.M. Kozlov, and E.S. Shmelev, Numerical study and optimization of the porous media VOC oxidizer with electric heating elements, *Int. J. Heat and Mass Transfer*, **49**, 5062-5069 (2006).

[9]I.Z. Ismagilov, E.M. Michurin, O.B. Sukhova, L.T. Tsykoza, E.V. Matus, M.A. Kerzhentsev, Z.R. Ismagilov, A.N. Zagoruiko, E.V. Rebrov, M.H.J.M. de Croon, and J.C. Schouten, Oxidation of organic compounds in a microstructured catalytic reactor, *J. Chem. Engr.*, **135S**, S57-S65 (2008).

[10]R.E. Hayes and S.T. Kolaczkowski, *Introduction to catalytic combustion*, Gordon and Breach Science Publishers, Netherlands (1997).

[11]L. Li, Z. Liu, Y. Qin, Z. Sun, J. Song, and L. Tang, Estimation of volatile organic compound mass transfer coefficients in the vacuum desorption of acetone from activated carbon, *J. Chem. Eng. Data*, **55**, 4732-4740 (2010).

[12]C.T.H. Stoddart and C. Kemball, The Catalytic Hydrogenation of Acetone on Evaporated Metallic Films, *J. Colloid Sci.*, **11**, 532-542 (1956)

[13]J. Gong, D.W. Flaherty, T. Yan, and C.B. Mullins, Selective Oxidation of Propanol on Au(111): Mechanistic Insights into Aerobic Oxidation of Alcohols, *ChemPhysChem*, **9**, 2461-2466 (2008).

[14]D. Bullen, X. Wang, J. Zou, S.-W. Chung, C.A. Mirkin, and C. Liu, Design, fabrication, and characterization of thermally actuated probe arrays for dip pen nanolithography, *J. MEMS*, **13**, 594-602 (2004).

[15]G. Langer, J. Hartmann, and M. Reichling, Thermal conductivity of thin metallic film measured by photothermal profile analysis, *Rev. Sci. Instrum.*, **68**, 1510-1513 (1997).

[16]J.H. Jou, C.N. Liao, and K.W. Jou, A method for the determination of gold thin-films mechanical-properties, *Thin Solid Films*, **238**, 70-72 (1994).

[17]A. Khan, J. Philip, and P. Hess, Young's modulus of silicon nitride used in scanning force microscope cantilevers, *J Appl. Phys.*, **95**, 1667-1672 (2004).

[18]V.K. Pamula, A. Jog, and R.B. Fair, Mechanical property measurement of thin-film gold using thermally acutated bimetallic cantilever beams, *Nanotech 2001*, **1**, 410-413 (2001).

[19]A. Piccirillo and A.L. Gobbi, Physical-Electrical Properties of Silicon Nitride Deposited by PECVD on I II-V Semiconductors, *J. Electrochem. Soc.*, **137**(12), 3910-3917 (1990).

[20]J.W.C. de Vries, Temperature and thickness dependence of the resistivity of thin polycrystalline aluminum, cobalt, nickel, palladium, silver and gold films, *Thin Solid Films*, **167**, 25-32 (1988).

[21]W.C. Gardiner, *Gas-phase combustion chemistry*, 1st Ed., Springer, New York (2000).

[22]NIST Chemistry Web Book (*http://webbook.nist.gov/chemistry*)

POLYANILINE-SILICA NANOCOMPOSITE: APPLICATION IN ELECTROCATALYSIS OF ACETYLTHIOCHOLINE

Prem C. Pandey*, Vandana Singh and S. Kumari
Department of Applied Chemistry, Institute of Technology, Banaras Hindu University,
Varanasi-221005, India

ABSTRACT

Electropolymerization of aniline is reported within nano-structured network of organically modified sol-gel glass (ORMOSIL) matrix. Three important redox mediators viz., tetracyanoquinodimethane (TCNQ), tetrathiafulvalene (TTF) and ferrocene were encapsulated within ormosil film at the surface of indium tin oxide (ITO) electrodes and used for electropolymerization of aniline. In order to study the effect of electrocatalyst together with the redox-mediator during the electropolymerization of aniline, palladium was also introduced within nano-structured network of ormosils. Polyaniline (PAni) with excellent electrochemical behavior was grown within these modified electrodes and the process of electropolymerization was found as a function of redox mediators' characteristics. The presence of palladium within ormosil matrix dramatically altered the electropolymerization process and the electrochemistry of PAni as well. The resulting polymers were analyzed by cyclic voltammetry and scanning electron microscopy and the results again provided remarkable dependence on the property of redox mediators and palladium content justifying similar conclusion. The polyaniline obtained as PAni-TCNQ and PAni-TCNQ-Pd composites has been utilized for fabricating the modified electrodes to study the electrochemical sensing of acetylthiocholine. The results based on cyclic voltammetry and amperometry justify that the electrode material exhibit excellent electrocatalytic activity for the oxidation of acetylthiocholine with major findings as compared to control: (1) acetylthiocholine undergo direct oxidation with considerable increase in both anodic and cathodic peak currents and (5) an increase in the sensitivity of acetylthiocholine analysis to the order of 5 fold for the modified electrodes.

INTRODUCTION

Polyaniline has been widely studied because of its potential applications in electrorheological fluids[1, 2], sensors[3], electrostatic discharge[4], and anticorrosion coatings[5]. Accordingly several routes of aniline polymerization in both aqueous and non-aqueous solvents with varying media compositions are available[6-9].

The electrochemical oxidation of aniline has typically been carried out in aqueous medium galvanostatically[10, 11], potentiostatically[6, 7, 12], or through cycling of the potential (potentiodynamic) of the substrate anode between suitable potential range versus Ag/AgCl or saturated calomel electrode (SCE)[7-9, 13, 14]. We have studied in details the potentstatic and potentiodynamic mode of electropolymerization of aniline[6, 7] in which several conventional electron transfer mediators showed reversible electrochemistry. In solution, the coupling of reversible redox electrochemistry together electrochemistry of aniline polymerization is difficult to observe due to fast dynamics of polymeric intermediates and redox couple of the electron transfer relays. On the other hand if redox mediators could be confined within nano-geometry of solid-state matrix i.e., organically modified sol-gel glass matrix (ORMOSIL) within which electropolymerization of aniline is triggered, it might be feasible to record such finding if the redox mediators are stable within selected potential range.

Earlier studies on the formation of conducting polymers together sol-gel glasses have been reported by Cox et al[15, 16] and others[17-23] following three different approaches. In first one, chemically prepared polymers were dissolved and mixed with a sol that was subsequently processed into a solid material[17, 18]. Second, the silica precursor was organically modified with a monomer; subsequent to the gelation, the polymerization was performed either chemically or electrochemically [19-21]. The third general method was to form a thin film of silica on an indium tin oxide electrode, immerse the system in a solution that contains the monomer, and perform the polymerization electrochemically[22, 23].From the past few years, efforts have been made to combine the physical strength, chemical stability, and optical properties of certain sol-gels with the electrical properties of conducting polymers for wide range of electrochemical applications. Additionally, encapsulation of electron transfer mediators[17-21] having reversible redox electrochemistry within ormosil films might facilitate charge-transfer process required for probing chemical / biochemical interaction taking place within or outside the nano-structured domains of solid-state and may prove as potential template for the electropolymerization of aniline. Earlier studies on the formation of PAni together sol-gel glasses have been reported by Cox *et al* [22, 23] and others[24-29] following three different approaches. In the first one, chemically prepared polymers were dissolved and mixed with a sol that was subsequently processed into a solid material[24-26]. Second, the silica precursor was organically modified with a monomer; subsequent to the gelation, the polymerization was performed electrochemically[26]. The third general method was to form a thin film of silica on an indium tin oxide electrode, immerse the system in a solution that contains the monomer, and perform the polymerization electrochemically[22,27].

The polymers synthesized under these conditions introduce electrocatalysis when used in electrochemical sensor's design. Further, many amperometric electrochemical sensors incorporate the participation of redox mediator for availing selective and sensitive detection of targeted analytes. Accordingly, the redox- mediated synthesis of PAni involving the participation of electron transfer relays may lead to promising materials for electrochemical sensors development. Additionally if the redox couple of the mediator has affinity toward the oxidation products of aniline, the resulting material may add facilitated electrocatalysis during sensing process. Such investigation has not been studied in sol-gel glasses manipulated with an electron transfer mediator together with metal catalyst which has been undertaken in this contribution. Here in this manuscript the electrocatalytic properties have been utilized to study the oxidation of acetylthiocholine (ATC).

Determination of acetylthiocholine (ATC) is of great interest since one of the products of acetylcholinesterase (AChE) mediated hydrolysis of acetylthiocholine is thiocholine and its detection can be used to assess the activity of AChE, a biomarker of the effect of pesticides (organophosphates (OPs) and carbamates) which inhibit cholinesterases[28]. Analysis of acetylthiocholine is, therefore, of great importance, particularly in the development of electrochemical sensors for detection of environmental pollutants such as OPs and carbamates[29].

Thus, in the present article, we report the electropolymerization of aniline within organically modified sol-gel glass (ORMOSIL) matrix derived through sol-gel process. It has been observed that the rate of electropolymerization was very slow within such matrix with poor electroactivity of the resulting polymer. Accordingly, it was planned to incorporate redox mediators along with metal catalyst within the ormosil matrix to enhance the rate of charge transfer process during electropolymerization of aniline. And finally utilizes the electrocatalytic properties of these materials in the electrochemical detection of acetylthiocholine.

EXPERIMENTAL

Materials

Aniline, 3-Aminopropyltrimethoxysilane, 3-Glycidoxypropyltrimethoxysilane, Tetracyanoquinodimethane (TCNQ), Tetrathiafulvalene (TTF), Ferrocene (Fc), Palladium chloride, graphite powder (particle size 1–2 μm) and Nujol oil (density 0.838) were obtained from Aldrich Chemical Co. 2-(3, 4-Epoxycyclohexyl) ethyltrimethoxysilane was obtained from Fluka. Acetylthiocholine was purchased from Sigma. All other chemicals employed were of analytical grade. Aniline was distilled under vacuum prior to use. The aqueous solutions of acetylthiocholine was freshly prepared in triply-distilled water before each experiment. Experiments were performed at room temperature.

Ormosil Films Preparations

A typical ormosil film was prepared by adding alkoxysilane precursors, TCNQ, TTF, Fc, hydrochloric acid and distilled water in the composition shown in Table 1. The mixture was vigorously stirred for 5 min. An aliquot of 10 μL of the suspension was layered on the indium tin oxide (ITO with surface resistance ~ 30 Ω) electrodes, and the electrodes were air-dried for 8-10 hours to ensure complete hydrolysis and gelation resulting into ormosil film.

Electrochemical Synthesis of Polyaniline Over Modified Electrodes

Polyaniline was synthesized electrochemically using single compartment cell equipped with three electrodes viz. ITO plate as working electrode, Ag/AgCl (Orion, Beverly, MA, USA) as reference electrode and Pt plate as the counter electrode. All the electrochemical work was done with an Electrochemical Workstation Model 660B, CH Instruments Inc., TX, USA. Polyaniline was deposited potentiodynamically over ormosil modified ITO electrodes from 1 M HCl with typical concentration of 0.1 M aniline by cycling the potential between -0.2 to 1.0 V versus Ag/AgCl. Electrochemical characterization of PAni films deposited over ITO electrodes was performed through cyclic voltammety by cycling the potential between -0.2 to 1.0 V versus Ag/AgCl in 1 M HCl at various scan rates viz. 0.01, 0.02, 0.05, 0.10 and 0.20 V/s. SEM studies of PAni films were performed using a FEI Quanta 200F Scanning electron microscope.

Modification of Graphite Paste Electrodes Through Composite Material

Polyaniline synthesized within TCNQ and TCNQ-Pd encapsulated ormosil matrix was extracted from ITO electrodes and suspended in tetrahydrofuran (5 mg in 1 ml THF) followed by sonication for 15 minutes. Then the resulting solution was then kept in oven at 60 ^{0}C for 1 hr. After complete evaporation of THF the dried material was used to modify graphite paste electrodes. The electrode body used for the construction of modified electrode was obtained from Bioanalytical Systems (West Lafayette, IN; (MF 2010)). The well was filled with an active paste of composition given in Table 2 (paste-1, paste-2 and paste-3). The desired amount of PAni-TCNQ and PAni-TCNQ-Pd composites was thoroughly mixed with graphite powder (particle size 1–2 μm) in a blender followed by addition of Nujol oil. After homogenization the mixture was stored into stoppered glass vial at room temperature when not in use. The paste surface was manually smoothened on a clean butter paper.

RESULTS AND DISCUSSION

Role of Nano-Structured Domains of Ormosil During Electropolymerization of Aniline

The current research program concerns to justify the role of nano-structured network, during electropolymerization of aniline. The nano-structured network may be readily introduced/controlled within organically modified silicate (ORMOSIL) matrix. Accordingly, it was planned to understand the process of electropolymerization on the surfaces of bare ATO electrode and ormosil-modified ATO electrode without redox mediator under similar experimental conditions. The results are shown in Fig.1 and they lead to following conclusions; (i) the process of electropolymerization is relatively much faster on bare electrode surface as compared to that on ormosil-modified electrode, (ii) the number of potentiodynamic cycles for detectable polymer properties on bare ITO is 20 whereas the same on ormosil-modified electrode is 100; (iii) The redox behavior of the polymer made on bare ATO surface is much poorer (Fig.1a) as compared that of ormosil-modified electrode (Fig.1b). These findings suggested that incorporation of nano-structured domains during electropolymerization process generates patterned polymeric structure with excellent redox behavior basically due to controlled growth of polyaniline domains within nano-structured network. Further, the effect of pore size over the polymerization of aniline within the silica matrix has been performed in detail by Cox et al[16] and demostrated that the size of the pores apparently limited the polymerization.

Role of Redox-Mediators And Metal Catalyst in the Electropolymerization of Aniline

Since the sol-gel glass itself is not an intrinsic conductor, the primary redox process, i.e., the oxidation of aniline, does not occur at the boundary between the outer layer of the silica and the liquid phase. The residual water with its dissolved acidic electrolyte that is present in sol-gel pores provides an environment for the coupled chemical reactions such as polymer formation inside the sol-gel pores; although poor conductivity under such condition may delay the polymer growth. This is well supported with the result of polymerization in sol-gel without mediator since the resulting polymer in this case took one hundred cycles to exhibit good redox behavior. Thus, to enhance the rate of polymerization of aniline and to understand the effect of redox mediators, PAni was grown within ormosil matrix encapsulating three-types of redox-transfer relays that essentially differ in their hydrophobicity/hydrophilicity. The organic redox mediators were TCNQ and TTF, well known redox components of organic metal and ferrocene was used as organometallic redox mediator. The results recorded on electropolymerization of aniline within ormosil network encapsulating TCNQ, TTF and Fc are shown in Fig. 2.

It is important here to discuss the role of ormosil-encapsulated redox mediators during electropolymerization of aniline. In order to understand such event the dynamics of ormosil-encapsulated redox mediator becomes crucial. The rotational degree of freedom of redox mediators within the ormosil matrix is fixed. Hence, the homogeneous mediation associated to oxidation of aniline monomers is not possible in the matrix. However, the results on electropolymerization of aniline are greatly influenced by the presence of ormosil-encapsulated mediatiors suggesting that the mediators under present conditions are acting as electrocatalyst during the polymerization of aniline. Additionally, palladium was also added to the network of sol-gel matrix to understand its effect on the polymerization process via interaction of palladium chloride and 3-glycidoxypropyltrimethoxysilane. The glycidyl group of 3-Glycidoxypropyltrimethoxysilane is highly reactive. When aqueous solution of palladium chloride, which acts as Lewis acid, is added, it opens the epoxide ring of the glycidyl moieties and in turn palladium is reduced. The reduced palladium was found coordinated with carbon atoms of glycidyl residue which were initially bonded to epoxide linkage[24, 25]. This reduced palladium acts as an electrocatalyst in the process of electropolymerization of aniline. The voltammograms shown in Fig.2 were recorded for 35 cycles without palladium and 35, 70 and 70 cycles for the systems with palladium (Fig. 3). This justifies two different conclusions based on the absence and the presence of palladium within nano-structured network. In the absence of palladium all three types of ormosil matrices resulted into three major characteristic redox peaks of polyaniline within 35 cycles whereas with palladium dramatic variation took place in the number of cycles required to obtain similar redox behavior of polyaniline. In case of palladium encapsulated ormosil, all systems except TCNQ showed increase in the number of cycles required to obtain similar redox behavior of PAni. Being a strong Π-acceptor ligand TCNQ have the capability to coordinate to metal ions and polymers both as neutral molecules and radical anions[30, 31] and thus forming a stable charge transfer complex. This special property of TCNQ might be responsible for all the anomaly observed during the electropolymerization of PAni over ormosil matrix.

Electrochemical Characterizations of PAni Grown Within Ormosil Matrix

The next stage of investigation is to understand the variable electrochemistry of PAni grown within these ormosil-modified electrodes. Cyclic voltammograms of PAni synthesized within ormosil matrices in absence of palladium at various scan rates are shown in Fig. 4, whereas the same with palladium are shown in Fig. 5. Results leads to following conclusions: (a) The anodic and cathodic peak currents are in the order of ferrocene > TTF > TCNQ; (b) the anodic and cathodic currents increase after adding palladium within ormosil network in case of ferrocene and TCNQ whereas a decrease takes place in case of TTF; (c) TCNQ-enapsulated ormosil showed relatively much better redox behavior of PAni in absence of palladium content as compared to all other systems; (d) redox electrochemistry of PAni after adding palladium content are relatively much better as compared to that of before adding palladium within ormosil network. These observations could be justified from the fact; (i) ferrocene is relatively much efficient mediator as compared to that of TTF and TCNQ; (ii) the anodic currents of PAni are the function of the efficiency of redox mediators, (iii) TCNQ seems to be most compatible mediator within ormosil matrix for rapid growth of PAni due to hydrophobicity and anionic behavior that suitably make charge transfer complex with available moieties within ormosil matrix.

Morphology of PAni Synsized Through Different Ormosil Matrices

The electrochemical polymerization provides the possibility of controlling the thickness and homogeneity of the polymers. Fig. 6 shows the typical SEM images of PAni grown electrochemically within silica matrix encapsulating different redox mediators. The images clearly depict that the nanostructure domains present in matrix of ormosil acted as a template for the synthesis of polymer. In all the cases, uniform interconnected fibrillar network of PAni was seen but in case of TCNQ more dense pattern and an ordered geometry of network of PAni was formed again confirms the interaction of TCNQ with PAni states leading to better growth of polymer chains.

Electrochemical Oxidation of Acetylthiocholine Over PAni-TCNQ Modified Electrodes

Acetylthiocholine (ATC) is electrochemically hydrolyzed into acetic acid and thiocholine amongst which thiocholine is electroactive. Direct hydrolysis of ATC over graphite paste electrode was previously studied in detail[32]. The cyclic voltammograms of different paste electrodes in the presence of 2 mM of ATC at the scan rate of 2 mV/ sec in phosphate buffer (100 mM, pH=7.0) are shown in Fig. 7. The corresponding anodic and cathodic peak currents, when the potential is in between -0.2 to 0.6 V versus Ag / AgCl, are found to be 14 µA / -9 µA, 25 µA / -12 µA and 37 µA / 18 µA for paste-1, paste-2 and paste-3 modified electrodes respectively. The result shows that, both oxidation and reduction currents are greater in paste-2 and paste-3 as compare to paste-1 suggests fast kinetics of spontaneous hydrolysis of ATC in the presence of electrocatalytic materials. In other words the process of hydrolysis facilitates in presence of PAni-TCNQ and PAni-TCNQ-Pd as compare to bare graphite paste electrode.

Amperometric determination of ATC was obtained by addition of different concentration (0.05 mM to 10 mM) of ATC in the phosphate buffer (100 mM, pH= 7.0) at working potential of 0.4 V. Fig. 8 shows the typical amperometric response curve for the three systems. It is very much clear from the figure that the response was highest in case of paste-3 followed by paste-2 and paste -1 again suggesting the excellent electrocatalytic behavior of composite material towards the hydrolysis of ATC. The calibration curves for ATC detection by amperometry at graphite paste electrodes modified with paste-1, paste-2 and paste-3 were constructed using average currents recorded at three individual electrodes for each concentration point. Fig. 9 shows the calibration curves for ATC for paste-1 (curve a), for paste-2 (curve b) and for paste-3 (curve c). The sensitivities towards ATC was found to be 0.14 $\mu A \pm 3$ nA/mM for paste-1 (curve a), 0.34 $\mu A \pm 6$ nA/mM for paste-2 (curve b) and 0.67 $\mu A \pm 8$ nA/mM for paste-3 (curve c) modified electrodes. The inset of Fig. 9 shows the linear range for ATC detection from 50 μM- 3 mM. A comparison on the performance of present sensor with earlier reported analogous systems has been given in Table 3. Since acetylthiocholine is enzymatically hydrolyzed into electroactive material thiocholine accordingly, many of the authors have studied acetylcholinesterase (AChE)-mediated detection of ATC. The non-enzymatic hydrolysis of ATC conducted in the present investigation at relatively lower operating potential with comparable sensitivity justifies the advantage of the present system over earlier reports (Table 3).

CONCLUSIONS

Electropolymerization of aniline within porous network of organically modified sol-gel glass (ormosil) encapsulating redox mediators is reported in the present work. Different types of redox mediators viz., TCNQ, TTF, Fc and an electrocatalyst palladium were encapsulated within the ormosil film to investigate their role in the rate of polymerization and electroactivity of PAni films formed over modified electrodes. The results obtained in this study indicate that the electrochemical performance of PAni is improved in the presence of redox mediator encapsulated within ormosil film. The encapsulated redox mediators act as a continuous conducting relay within the porous ormosil matrix, thereby providing an electronic conduction pathway which improves the process of charge transfer through the matrix. Such matrix apparently behaves as an electrocatalyst for aniline oxidation. The introduction of the known electrocatalyst together with redox mediators further improves the electroactivity of the polymer. Further, the application of these materials was investigated in the electrocatalytic oxidation of acetylthiocholine and it was concluded that resulting composite materials of PAni with TCNQ and TCNQ-Pd show excellent electrocatalytic behavior towards the oxidation of acetylthiocholine.

ACKNOWLEDGEMENT

The authors are thankful to CSIR, New Delhi for financial support.

REFERENCES

[1] M.S. Cho, Y.H. Cho, H.J. Choi and M.S. Jhon, Synthesis and Electrorheological Characteristics of Polyaniline-Coated Poly(methyl methacrylate) Microsphere: Size Effect, *Langmuir*, **19**, 5875-81 (2003).

[2] K.Aoki, J Chen, Q.Ke, S.P.Armes and D.P. Randall, Redox Reactions of Polyaniline-Coated Latex Suspensions, *Langmuir*, **19**, 5511-16 (2003).

[3] J.Huang, S.Virji, B.H Weiller and R.B. Kaner, Polyaniline Nanofibers: Facile Synthesis and Chemical Sensors, *J. Am. Chem. Soc.*, **125**, 314-15 (2003).

[4] V.G. Kulkarni, Tuned conductive coatings from polyaniline,*Synth. Met.*, **71**, 2129-31 (1995).

[5] B. Wessling and J. Posdorfer, Nanostructures of the dispersed organic metal polyaniline responsible for macroscopic effects in corrosion protection, *Synth. Met.*, **102**,1400-01 (1999).

[6] P.C. Pandey and G. Singh, Tetraphenylborate doped polyaniline based novel pH sensor and solid-state urea biosensor, *Talanta*, **55**, 773-82 (2001).

[7] P.C. Pandey and G. Singh, Electrochemical Polymerization of Aniline in Proton-Free Nonaqueous Media Dependence of Microstructure and Electrochemical, *J. Electrochem. Soc.*, **149**, D51-D56 (2001).

[8] D.D. Borole, U.R. Kapadi, P.P. Kumbhar and D.G. Hundiwale, Oxidation behavior of micro- and nano-crystalline coatings deposited by series double-pole electro-pulse discharge, *Mater. Lett.*, **56**, 85-92 (2002).

[9] J.R. Santos Jr, J.A. Malmonge, A.J.G. Conceicgo Silva, A.J. Motheo, Y.P. Mascarenhas and L.H.C. Mattoso, Characteristics of polyaniline electropolymerized in camphor sulfonic acid, *Synth. Met.*, **69**, 141-42 (1995).

[10] Z.J. Ling, Z.X. gang, X. Fang and H.F. Ping, Effect of polar solvent acetonitrile on the electrochemical behavior of polyaniline in ionic liquid electrolytes, *J. Colloid Interface Sci.*, **287**, 67-71 (2005).

[11] K. Luo, N. Shi and C. Sun, Thermal transition of electrochemically synthesized polyaniline, *Polym. Degrad. Stab.*, **91**, 2660-64 (2006).

[12] A.A. Nekrasov, V.F. Ivanov, O.L. Gribkova and A.V. Vannikov, Voltabsorptometric study of "structural memory" effects in polyaniline, *Electrochim. Acta*, **50**, 1605-13 (2005).

[13] R. Prakash, Electrochemistry of Polyaniline: Study of the pH Effect and Electrochromism, *J. Appl. Polym. Sci.*, **83**, 378-85 (2002).

[14] A.T. Özyılmaz, M. Erbil and B. Yazıcı, The electrochemical synthesis of polyaniline on stainless steel and its corrosion performance, *Current Appl. Phys.*, **6**, 1-9 (2006).

[15] J.Widera and J.A. Cox, Electrochemical oxidation of aniline in a silica sol–gel matrix, *Electrochem. Commun.*, **4**, 118-22 (2002).

[16] J. Widera, A.M. Kijak, D.V. Ca, G.E. Pacey, R.T. Taylor, H. Perfect and J.A. Cox, The influence of the matrix structure on the oxidation of aniline in a silica sol–gel composite, *Electrochim. Acta*, **50**, 1703-09 (2005).

[17] B.R Mattes, E.T. Knobbe, P.D. Fuqua, F. Nishida, E.-W. Chang, B.M. Pierce, B. Dunn and R.B. Kaner, Polyaniline sol-gels and their third-order nonlinear optical effects, *Synth. Met.*, **43**, 3183-87 (1991).

[18] Y. Wei, J.-M. Yen, D. Jin, X. Jia, J. Wang, G.-W. Jang, C. Chen and R.W. Gumbs, Composites of Electronically Conductive Polyaniline with Polyacrylate-Silica Hybrid Sol-Gel Materials, *Chem. Mater.*, **7**, 969-74 (1995).

[19] R.J.P. Corriu, J.J.E. Moreau, P. Thepot, M.W.C. Man, C. Chorro, J.-P. L_ere-Porte and J.-L. Sauvajol, Trialkoxysilyl Mono-, Bi-, and Terthiophenes as Molecular Precursors of Hybrid Organic-Inorganic Materials, *Chem. Mater.*, **6**, 640-49 (1994).

[20]C. Sanchez, B. Alonso, F. Chapusot, F. Ribot and P.J. Audebert, Molecular design of hybrid organic-inorganic materials with electronic properties, *J. Sol-Gel Sci. Technol.*, **2**, 161-66 (1994).

[21]G.-W. Jang, C. Chen, R.W. Gumbs, Y. Wei and J.-M. Yeh, Composites of Polyaniline and Polyacrylate-Silica Hybrid Sol-Gel Materials, *J. Electrochem. Soc.,***143**, 2591-96 (1996).

[22]M.M. Verghese, K. Ramanathan, S.M. Ashraf, M.N. Kamalasanan and B.D. Malhotra, Electrochemical Growth of Polyaniline in Porous Sol-Gel Films, *Chem. Mater.*, **8**, 822-24 (1996).

[23]S. das Neves, S.I. C_ordoba de Torresi and R.Ap. Zoppi, Template synthesis of polyaniline: a route to achieve nanocomposites, *Synth. Met.*, **101**, 754-755 (1999).

[24]P.C.Pandey, S. Upadhyay and S. Sharma, Functionalized Ormosils-Based Biosensor Probing a Horseradish Peroxidase-Catalyzed Reaction, *J. Electrochem. Soc.,***150**, H85-H92 (2003).

[25] P.C. Pandey, S. Upadhyay, I. Tiwari and S. Sharma, You have full text access to this contentA Novel Ferrocene-Encapsulated Palladium-Linked Ormosil-Based Electrocatalytic Biosensor. The Role of the Reactive Functional Group, *Electroanalysis*, **13**, 1519-27 (2001).

[26]G.-W. Jang, C. Chen, R.W. Gumbs, Y. Wei and J.-M. Yeh, Large-Area Electrochromic Coatings, *J. Electrochem. Soc.*, **143**, 2591-96 (1996).

[27]Y. Xian, F. Liu, L. Feng, F. Wu, L. Wang and L. Jin, Nanoelectrode ensembles based on conductive polyaniline/poly(acrylic acid) using porous sol–gel films as template, *Electrochem. Commun.*, **9**, 773-80.(2007).

[28]H. Schulze, S. Vorlov'a, F. Villatte, T.T. Bachmann and R.D. Schmid, Design of acetylcholinesterases for biosensor applications, *Biosens. Bioelectron.*, **18**, 201-09 (2003).

[29]E. Suprun, G. Evtugyn, H. Budnikov, F. Ricci, D. Moscone and G. Palleschi, Acetylcholinesterase sensor based on screen-printed carbon electrode modified with prussian blue, *Anal. Bioanal. Chem.*, **383**, 597- 604 (2005).

[30]N. Mincheva, L. Ballester, L. Antonov and M. Mitewa, New Dimeric Pd(III) Pd(II) Complex with 7,7,8,8-Tetracyanoquinodimethane (TCNQ), *Synth. React. Inorg. Met.-Org. Chem.*; **30**, 1643-51 (2000).

[31]J.C. Li, Q. Xue, Y. Zeng, W.M. Liu, Q.D. Wu, Y.L. Song, and L. Jiang, Growth and characterization of polyaniline 7,7,8,8-tetracyanoquino-dimethane (TCNQ) complex films grown by vacuum evaporation, *Thin Solid Films*, **374**, 59-63 (2000).

[32]P.C. Pandey, S. Upadhyay, H.C. Pathak, C.M.D. Pandey and I. Tiwari, Acetylthiocholineracetylcholine and thiocholinercholine electrochemical biosensorsrsensors based on an organically modified sol–gel glass enzyme reactor and graphite paste electrode, *Sens. Actuators, B*, **62**, 109-116 (2000).

[33]V. S. Somerset, M. J. Klink, P.G. L. Baker and E. I. Iwuoha, Acetylcholinesterase-polyaniline biosensor investigation of organophosphate pesticides in selected organic solvents, *J. Environ. Sci. Health., Part B*, **42**, 297-304 (2007).

[34]O. Shulga and J. R. Kirchhoff, An acetylcholinesterase enzyme electrode stabilized by an electrodeposited gold nanoparticle laye, *Electrochem. Commun.*, **9**, 935-40 (2007).

[35]D. Du, X. Huang, J. Cai, A. Zhang, J. Ding and S. Chen, An amperometric acetylthiocholine sensor based on immobilization of acetylcholinesterase on a multiwall carbon nanotube–cross-linked chitosan composite, *Anal Bioanal. Chem.*, **387**, 1059- 65 (2007).

Table-1: Composition of different ormosil-modified electrodes

System	A* (μl)	B* (μl)	C* (μl)	D* (μl)	E* (μl)	F* (μl)	G* (μl)	H* (μl)	I* (μl)
Blank sol gel	70	10	-	-	-	-	-	5	300
TCNQ	-	10	70	-	-	-	-	5	300
TCNQ-Pd	-	10	70	-	-	10	10	5	280
TTF	-	10	-	70	-	-	-	5	300
TTF-Pd	-	-	-	70	-	10	10	5	280
Fc	-	-	-	-	70	-	-	5	300
Fc-Pd	-	-	-	-	70	10	10	5	280

A* = 3-Aminopropyltrimethoxysilane, B* = 2-(3, 4-epoxycyclohexyl) ethyltrimethoxysilane, C* = TCNQ in 3-Aminopropyltrimethoxysilane (45 mM), D* = TTF in 3-Aminopropyltrimethoxysilane (45 mM), E* = Fc in 3-Aminopropyltrimethoxysilane (45 mM), F* = 3-Glycidoxypropyltrimethoxysilane, G* = Palladium Chloride (3mg/mL), H* = 0.1 M HCl, I* = Distilled water

Table- 2: Composition of graphite paste electrodes

Systems	PAni-TCNQ / PAni-TCNQ-Pd (% w/w)	Graphite powder (% w/w)	Mineral oil (% w/w)
Paste-1	-	70	30
Paste-2	1	69	30
Paste-3	1	69	30

Table-3: Comparison on the performance of some modified electrodes used in the electrocatalysis of acetylthiocholine

Substrate	Modifier	pH	Operating potential (mV)	Stability	Reference
Au electrode	PAni with AChE	7.2 (phosphate buffer)	+400	Poor	[33]
CP electrode	AChE encapsulated Ormosil	6 & 8 (Tris-HCl buffer)	+350	Good	[32]
Au electrode	AChE immobilized over AuNps	8 (phosphate buffer)	+680	Poor	[34]
GC electrode	AChE covalent bonded with MWNT cross-linked chitosan composite	7 (phosphate buffer)	+800	Good	[35]
CP electrode	PAni-TCNQ & PAni-TCNQ-Pd **(without enzyme)**	7 (phosphate buffer)	+400	Good	Present work

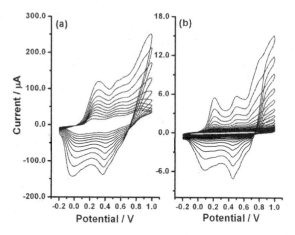

Fig. 1. Potentiodynamic electropolymerization of 0.1 M aniline in 1.0 M HCl at the scan rate of 0.05 Vs^{-1} over ITO modified with (a) bare ITO; (b) blank sol-gel.

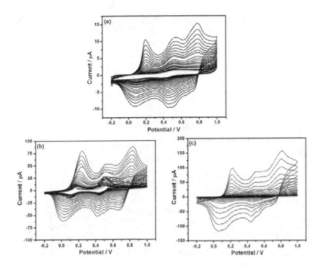

Fig. 2. Potentiodynamic electropolymerization of 0.1 M aniline in 1.0 M HCl at the scan rate of 0.05 Vs^{-1} over ITO modified with (a) TCNQ; (b) TTF and (c)Fc.

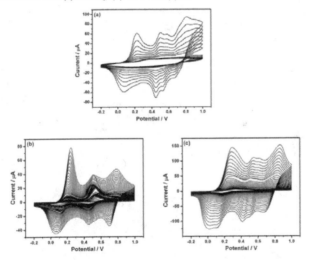

Fig. 3. Potentiodynamic electropolymerization of 0.1 M aniline in 1.0 M HCl at the scan rate of 0.05 Vs^{-1} over ITO modified with (a) TCNQ-Pd; (b) TTF-Pd and (c) Fc-Pd.

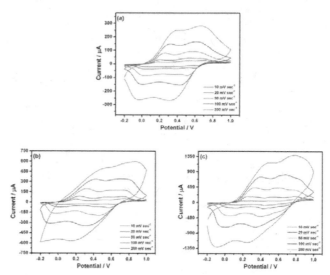

Fig. 4. Cyclic voltammogram of PAni in 1.0 M HCl over ITO modified with (a) TCNQ; (b) TTF and (c) Fc at different scan rates.

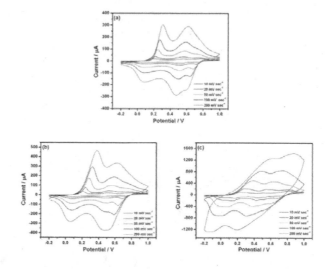

Fig. 5. Cyclic voltammogram of PAni in 1.0 M HCl over ITO modified with (a)TCNQ-Pd; (b) TTF-Pd and (c) Fc-Pd at different scan rates.

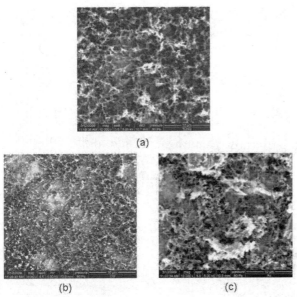

Fig. 6. SEM images of PAni synthesized within ormosil matrix encapsulated with (a) TCNQ; (b) TTF and (c) Fc.

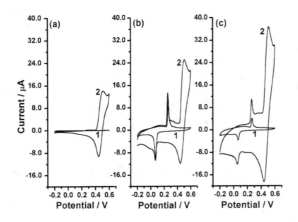

Fig. 7. Cyclic voltammograms of graphite paste electrodes modified with (a) (1) paste-1, (b) (1) paste-2 and (c) (1) paste-3; (2) + 2 mM ATC in 100 mM phosphate buffer (pH 7.0) at the scan rate of 0.002 Vs^{-1}.

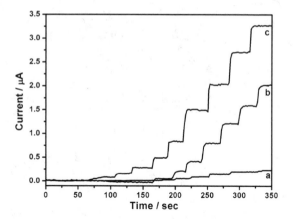

Fig. 8. Typical amperometric response curves on the addition of varying concentrations of ATC over graphite paste electrode made from (a) paste-1, (b) paste-2 and (c) paste-3 at 0.4 V vs. Ag / AgCl at 25 °C in 100 mM phosphate buffer (pH 7.0).

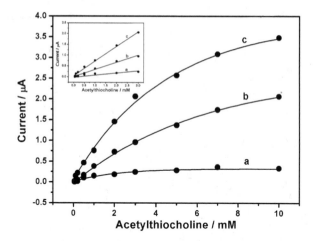

Fig. 9. Calibration curve for the analysis of ATC over graphite paste modified electrode with (a) paste-1, (b) paste-2 and (c) paste-3. The inset shows the linear relationship between anodic current and concentration of ATC.

ELECTROCHEMICAL SENSING OF DOPAMINE OVER POLYINDOLE-COMPOSITE ELECTRODE

Prem C. Pandey[*], Dheeraj S. Chauhan, and S. Kumari[1]
Department of Applied Chemistry, Institute of Technology, Banaras Hindu University
Varanasi-221005, Uttar Pradesh, India: [-1] Department of Chemistry, Banaras Hindu University, India

ABSTRACT

We hereby report a new conducting polymer composite i.e., polyindole-camphorsulphonic acid (PIn-CSA) composite, and its use in the development of dopamine sensor. The processable PIn-CSA composite is developed by homogenizing equimolar ratio of chemically synthesized polyindole and camphor sulphonic acid in tetrahydrofuran. The homogenized PIn-CSA composite is cast over the Pt disc electrode under ambient conditions and assembled in a homemade electrode body equipped with Ag/AgCl reference electrode. The ion sensor exhibits excellent response towards dopamine in presence of ascorbic acid over a wide concentration range. The sensor can be used for 1 month without any major drift in its sensitivity and limit of detection.

INTRODUCTION

The design of chemical sensors for selective detection of a specific analyte is a topic of considerable interest, due to their wide ranging application in the broad areas of chemistry and biology From the past few years, there has been a considerable effort in the development of voltammetric methods for the determination of DA and AA in biological samples[1-8]. The direct redox reactions of these species at bare electrodes are irreversible and therefore require high overpotentials[9]. Moreover, the direct redox reactions of these species at the bare electrodes take place at very similar potentials and often suffer from a pronounced fouling effect, which results in rather poor selectivity and reproducibility. Thus, to determine DA selectively in presence of AA has been a major goal of electroanalytical research. Many workers have utilized electropolymerized films of conducting polymers which are uniform and strongly adherent to the electrode surface. Conducting polymers, such as polyaniline, polypyrrole and poly (3-methylthiophene) have been reported for DA determination in an excess of AA[9]. The most extensively studied polymer for DA sensing is overoxidized polypyrrole[10,11] and its derivatives[12,13].

The study of the electrochemical properties of conducting polymers containing nitrogen atoms such as polyaniline, polypyrrole, polycarbazole and their substituted derivatives[14-16] has attracted considerable attention since these materials can be utilized in interesting applications, such as electrochromic devices[17], battery electrodes[18] and chemical sensors[15,18].

However, in comparison to other conducting polymers mentioned above the study of polyindole and its derivatives has only scarcely been investigated[19,20]. This is because polyindole, which is mainly synthesized by chemical oxidation or electropolymerization of indole monomer, has low polymerization efficiency[21,22]. But this polymer and its derivatives appear to be good candidates for applications in various fields like electronics and electrocatalysis.

We report here a new composite of polyindole with camphorsulphonic acid synthesized following chemical route and its application in the construction of dopamine sensor. This configuration showed advantage over similar sensors reported earlier based on the use of conducting polymer composites or a single conducting polymer sensing layer, which is directly in contact with transducer. The new dopamine ion sensor shows improved selectivity in the presence of 100 times more concentration of AA which has been one of the important requirements of such sensor for practical applications.

EXPERIMENTAL

Materials and Methods

Indole and ammonium peroxodisulphate were obtained from Merck, India. Nafion®, a 5 wt. % solution in a mixture of lower aliphatic alcohols and 20% water was obtained from Aldrich. Camphorsulponic acid (CSA) was obtained from HiMedia, India. Ascorbic acid (AA) and Dopamine (DA) were purchased from Sigma Chemicals and used as received. All other chemicals used were of analytical grade. Aqueous solutions were prepared with double distilled water. Freshly prepared AA and DA were used for all experiments. All the experiments were performed at room temperature.

Electrochemical experiments were performed on an Electrochemical Workstation Model CHI660B, CH Instruments Inc., TX, USA. Cyclic voltammetry of modified electrodes was performed with a working volume of 3 mL of 0.5 M sulphuric acid. An Ag/AgCl electrode (Orion, Beverly, MA, USA) and a platinum plate electrode served as reference and counter electrode, respectively. All potentials given below are relative to the Ag/AgCl reference electrodes. The working electrode in all the experiments was a modified platinum disc (2 mm diameter). The composite was cast over platinum electrode and cyclic voltammetry was carried out in 0.5 M H_2SO_4 for scan rates viz. 0.01, 0.02, 0.05, 0.10, 0.20 and 0.50 Vs^{-1} to study the electrochemical behavior. The UV–vis spectra of chemically synthesized polyindole (PIn) and that of the composite were recorded by dissolving them in dimethyl sulphoxide (DMSO) and tetrahydrofuran (THF) respectively using a Systronics UV-VIS Double Beam Spectrophotometer 2201.

Synthesis of PIn-CSA Composite

Polyindole was prepared by the oxidative polymerization of indole using ammonium peroxodisulphate as described earlier [30]. After purification of the polymer following washing and drying steps, the next stage was to mix the polymer with camphorsulphonic acid. PIn was mixed with CSA in a mortar pestle in an equimolar ratio. THF was added drop wise to dissolve PIn and CSA resulting into a dark brown coloured solution. This solution was stirred well and used for characterization as well as in the formation of composite films and sensor application.

Construction of Modified Electrodes

An aliquot of 3 µL of the PIn-CSA solution in THF was cast over Pt disc electrode by solution casting method. After solvent evaporation, the composite film was dried in air for 12 hrs. 3 µL Nafion was added to this electrode in order to prepare a PIn-CSA-Nafion-modified electrode. The electrode was dried in air for 2 hrs. The dried electrodes were used for electrochemical investigations of DA and AA.

Dopamine Voltammetry over Modified Electrodes

The modified electrodes were used for voltammetric detection of DA and AA in 100 mM phosphate buffer (pH 7.4). The cyclic voltammograms were recorded before and after the addition of 0.1 mM DA and 1 mM AA at the scan rate of 0.01 Vs^{-1} between the potential range of -0.2 to 0.6 V. Differential pulse voltammety (DPV) was performed over the modified electrodes to ascertain the limit of detection for DA sensing and sensitivity of the modified electrodes towards DA. DPV experiments were performed in a three electrode assembly from -0.2 V to 0.6 V with an amplitude of 0.05 V and pulse period of 0.2 s within a concentration range of 0.001 mM to 0.1 mM DA in presence of 1 mM AA in 100 mM phosphate buffer (pH 7.4).

RESULTS AND DISCUSSION

Cyclic Voltammetry of PIn-CSA Modified Electrode

Cyclic voltammograms for PIn-CSA were recorded in 0.5 M H_2SO_4 by modifying Pt disk electrode at various scan rates viz. 0.01, 0.02, 0.05, 0.10, 0.20 and 0.50 Vs^{-1} as shown in Figure 1. In both the cases the insets show the voltammograms obtained at 0.01 Vs^{-1}. PIn showed anodic peaks (Epa) at 0.60 and 0.90 V, and cathodic peaks (Epc) at 0.41 and 0.77 V corresponding to change in the state of polymer. In PIn-CSA, again there are two redox couples with anodic peaks (Epa) at 0.58 V and 0.84 V and cathodic peaks (Epc) at 0.34 and 0.74 V. But in this case there is a minor shift in the Epa and Epc values to the lower side. This indicates the formation of an electroactive composite of PIn with CSA. Also, it is quite clear that in PIn-CSA, the two redox couples can easily be distinguished even at higher scan rates indicating that the composite has retained its electroactive behaviour with respect to increasing scan rates.

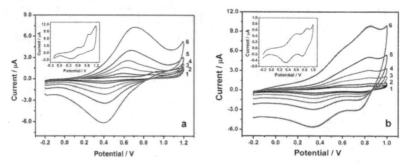

Figure 1. Cyclic voltammograms of (a) PIn and (b) PIn-CSA in 0.5 M H$_2$SO$_4$ at different scan rates viz. 0.01, 0.02, 0.05, 0.10, 0.20 and 0.50 Vs^{-1} (curves from lower to higher scan rates). Insets show the voltammograms at 0.01 Vs^{-1}.

UV–vis Studies

Figure 2 shows the UV-vis spectra of PIn and PIn-CSA. Characteristic peaks of PIn are also observed at 320, 367 and 398 nm. The spectrum of the composite is similar to that of PIn, However, a broad hump was observed at higher wavelength due to interaction of CSA with polymer chains.

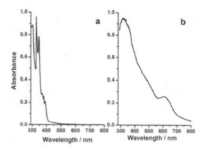

Figure 2. UV-vis spectra of (a) PIn and (b) PIn-CSA.

Cyclic Voltammetry of Dopamine over modified electrode

The response of DA along in presence of AA was investigated over Pt disc modified with PIn-CSA-Nafion and the results are shown in Figure 3. Curve a shows the behavior of PIn-CSA-Nafion-modified electrode. Curves b and c show the voltammograms after addition of 1 mM of AA and 0.1 mM of DA, respectively. It was observed that the oxidation of AA was suppressed, whereas the oxidation of DA was favored over the modified electrode. This result confirmed the role of polymer in the selective sensing of DA in presence of AA. The selectivity of modified electrode towards DA is attributed to the presence of anionic groups of CSA in the polymer backbone resulting in permselectivity for the cations of DA. Also, Nafion, a cation-exchange polymer, whose films are highly permeable to cations but almost impermeable to anions, contributes to eliminate the interference of AA with DA detection.

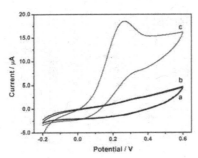

Figure. 3. Cyclic voltammogram of (a) Pt disc modified with PIn-CSA-Nafion film (b) + 0.1 mM AA (c) + 1 mM DA in 100 mM phosphate buffer (pH 7.4) at the scan rate of 0.01 Vs^{-1}.

Differential Pulse Voltammetry of Dopamine over modified electrode

The differential pulse voltammetric (DPV) response of varying concentrations of DA was recorded over modified electrode in presence of 1 mM AA and is shown in Figure 4. Almost negligible response could be observed when 1 mM AA was added into the working medium i.e., phosphate buffer solution. Also, with an increase in DA concentration (represented by the double headed arrow c) the peak current for PIn-CSA-Nafion sites also increased. The inset shows the DPV response of 0.1 mM DA (curve c) in presence of 1 mM AA (curve b) over modified electrodes. It is clear that no oxidation peak for AA is present in while a well resolved oxidation peak of DA at 100 times less concentration to that of AA occurs at 0.20 V, suggesting the selectivity towards DA over modified electrodes.

The calibration curves for DA detection by DPV at the modified electrodes were constructed using average currents recorded at three individual films for each concentration point. Figure 6 shows the calibration curves for DA detection in presence of 1 mM AA for PIn-CSA-Nafion-modified electrode. The sensitivity towards DA sensing was found to be 25 ± 0.007 μA per mM of DA concentration for the modified electrode. Further, the limit of detection for DA sensing was found to be 1.5 μM. The enhanced sensitivity and improved linear response for DA detection of PIn-CSA-Nafion-modified electrode proves its superiority over earlier work related to polyindole [5]. Such large amplification in DA sensing could be explained from following considerations. Nafion is also permselective for cations as PIn-CSA thus, the combination of selectivity of both facilitates the diffusion of DA cation into the polymer matrix.

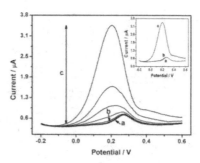

Figure 4. Differential pulse voltammograms of (a) GCE modified with PICA-Nafion film (b) + 1 mM AA (c) repeated additions of DA from 0.1 μM to 0.1 mM in presence of 1 mM AA in 100 mM phosphate buffer (pH 7.4). The inset shows DPV of (b) 1 mM AA (c) + 0.1 mM DA.

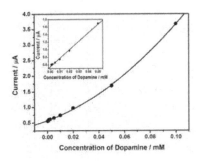

Figure 5. Calibration curve for the analysis of 0.1 μM to 0.1 mM DA over GCE modified with (a) PICA-Nafion (b) PICA-TCNQ-Nafion films. The inset shows the linear relationship between anodic current and concentration of DA.

Lifetime and Stability

The prepared electrode's analytical usefulness time was tested by measuring characteristic slopes i.e., sensitivity systematically, usually every 5 days. The drift in the limit of detection of the PIn-CSA-Nafion-modified electrode was also studied. This was found to be in the order of 3-4 μM in 30 days. In addition a small drift in the sensitivity shows the potential application of the composite film for the construction of DA sensor suitable for practical application. The life time for a typical sensor is shown in Table I.

Table I. Life time and drift in the response of a typical DA sensor

Days	Limit of Detection (μM)	Sensitivity (μA/mM)
1	1.5	25.0 ± 0.007
10	1.9	23.8 ± 0.009
20	3.0	22.9 ± 0.008
30	4.8	21.6 ± 0.007

CONCLUSIONS

The present investigation relates to the electrochemical sensing of dopamine (DA) over processable polyindole-camphor sulphonic acid (PIn-CSA) composite. The composite was synthesized following facile chemical method and characterized using various tools for its structural and electrochemical properties. The processable electroactive composite was utilized for the construction of a highly sensitive, low cost, user friendly dopamine sensor using platinum disc electrode. The sensor depicted an excellent response for dopamine in a wide range of concentration from 5 μM to 0.1 M with high stability and reproducibility with negligible drift of the standard potential. The sensitivity was calculated to be 25 ± 0.007 μA per mM of DA. Negligible interference was observed in presence of ascorbic acid (AA). The limit of detection for the DA sensor was found out to be 1.5 μM with a wide range of linearity.

ACKNOWLEDGEMENT

The authors gratefully acknowledge the Council of Scientific and Industrial Research (CSIR), India, for providing financial support to this work.

REFERENCES

[1] Pihel K., Walker Q.D., and Wightman R.M., Overoxidized polypyrrole-coated carbon fiber microelectrodes for dopamine measurements with fast-scan cyclic voltammetry, *Anal. Chem.*, 68, 2084–2089, 1996.

[2] Yan W., Feng X., Chen X., Li X., and Zhu J.-J., A selective dopamine biosensor based on AgCl@polyaniline core–shell nanocomposites, *Bioelectrochem.*, 72, 21–27, 2008.

[3] Ghita M., and Arrigan D.W.M., Dopamine voltammetry at overoxidised polyindole electrodes, *Electrochim. Acta*, 49, 4743–4751, 2004.

[4] Wang X., Yang N., Wan Q., and Wang X., Catalytic capability of poly(malachite green) films based electrochemical sensor for oxidation of dopamine, *Sens. Actuators, B*, 128, 83–90, 2007.

[5] Li Y., Wang P., Wang L., and Lin X., Overoxidized polypyrrole film directed single-walled carbon nanotubes immobilization on glassy carbon electrode and its sensing applications, *Biosens. Bioelectron.*, 22, 3120–3125, 2007.

[6] Feng X., Mao C., Yang G., Hou W., and Zhu J.-J., Polyaniline/Au composite hollow spheres: synthesis, characterization, and application to the detection of dopamine, *Langmuir*, 22, 4384–4389, 2006.

[7] Wang H.-S., Li T.-H., Jia W.-L., and Xu H.-Y., Highly selective and sensitive determination of dopamine using a Nafion/carbon nanotubes coated poly(3-methylthiophene) modified electrode, *Biosens. Bioelectron.* 22, 664–669, 2006.

[8] Pandey P.C., and Upadhyay B.C., Studies on differential sensing of dopamine at the surface of chemically sensitized ormosil-modified electrodes, *Talanta*; 67, 997–1006, 2005.

[9] H.T. Xu, F. Kitamura, T. Ohsaka, K. Tokuda, Anal. Sci. 10 (1994) 399.

[10] Z.Gao, B. Chen, M. Zi, Analyst 119 (1994) 459.

[11] A.M. Farrington, J.M. Slater, Electroanalysis 9 (1997) 843.

[12] J. Wang, P.V.A. Pamidi, G. Cerpia, S. Basac, K. Rajeshwar, Analyst 122 (1997) 981.

[13] D.W.M. Arrigan, Anal. Commun. 34 (1997) 241.

[14] Abe S., Kijima M., and Shirakawa H., Effect of mesogenic cores and length of spacers on liquid crystallinity of N-substituted polypyrrole derivatives, *Synth. Met.*, 119, 421–422, 2001.

[15] Athawale A.A., and Kulkarni M.V., Polyaniline and its substituted derivatives as sensor for aliphatic alcohols, *Sens. Actuators, B*, 67, 173–177, 2000.

[16] Lallemand F., Auguste D., Amato C., Hevesi L., Delhalle J., and Mekhalif Z., Electrochemical synthesis and characterization of *N*-substituted polypyrrole derivatives on nickel, Electrochim. Acta, 52, 4334–4341, 2007.

[17] Lim J.Y., Ko H.C., and Lee H., Single- and dual-type electrochromic devices based on polycarbazole derivative bearing pendent viologen, *Synth. Met.*, 156, 695–698, 2006.

[18] Lindfors T., and Ivaska A., pH sensitivity of polyaniline and its substituted derivatives, *J. Electroanal. Chem.*, 531, 43–52, 2002.

[19] Nie G., Zhou L., Guo Q., and Zhang S., A new electrochromic material from an indole derivative and its application in high-quality electrochromic devices, *Electrochem. Commun.*, 12, 160–163, 2010.

[20] Xu J., Zhou W., Hou J., Pu S., Yan L., and Wang J., Electrosyntheses of high quality poly (5-cyanoindole) films in boron trifluoride diethyl etherate containing additional diethyl ether, *Mater. Chem. Phys.*, 99, 341–349, 2006.

[21] Billaud D., Maarouf E.B., and Hannecart E., Chemical oxidation and polymerization of indole, Synth. Met., 69, 571–572, 1995.

[22] Saraji M., and Bagheri A., Electropolymerization of indole and study of electrochemical behavior of the polymer in aqueous solutions, *Synth. Met.*, 98, 57–63, 1998.

[23] B. Gupta, D.S. Chauhan, R. Prakash, Controlled morphology of conducting polymers: Formation of nanorods and microspheres of polyindole, Mater. Chem. Phys. 120, (2010) 625–630.

Author Index